DEPARTMENT OF DENTAL MATERIALS SCIENCE
U.M.D.S. DENTAL SCHOOL, GUY'S HOSPITAL
LONDON, SE1 9RT

The Application of Shape Memory Alloys in Medicine

The Application of Shape Memory Alloys in Medicine

by

I P Lipscomb

and

L D M Nokes

First published 1996

This publication is copyright under the Berne Convention and the International Copyright Convention. All rights reserved. Apart from any fair dealing for the purpose of private study, research, criticism, or review, as permitted under the Copyright Designs and Patents Act, 1988, no part may be reproduced, stored in a retrieval system, or transmitted in any form or by any means, electronic, electrical, chemical mechanical, photocopying, recording or otherwise, without the prior permission of the copyright owners. Unlicensed multiple copying of this publication is illegal. Inquiries should be addressed to: The Publishing Editor, Mechanical Engineering Publications Limited, Northgate Avenue, Bury St Edmunds, Suffolk, IP32 6BW, UK.

© I P Lipscomb and L D M Nokes

ISBN 0 85298 956 3

A CIP catalogue record for this book is available from the British Library.

The publishers are not responsible for any statement made in this publication. Data, discussion, and conclusions developed by the Authors are for information only and are not intended for use without independent substantiating investigation on the part of the potential users. Opinions expressed are those of the Authors and are not necessarily those of the Institution of Mechanical Engineers or its publishers.

Typeset by Paston Press Ltd, Loddon, Norfolk
Printed in Great Britain by Antony Rowe Ltd, Chippenham, Wiltshire

Contents

Foreword   vii

Preface   ix

Acknowledgements   xi

Chapter 1   Introduction to shape memory alloys (SMAs)   1
    1.1   The history of shape memory alloys   1
    1.2   The mechanics of the shape memory effect   1
    References   3

Chapter 2   Characteristics of shape memory alloys in medical applications   5
    2.1   Biocompatibility of Ni–Ti alloy   5
       2.1.1   *In vivo* testing of Ni–Ti alloy   5
       2.1.2   *In vitro* tesing of Ni–Ti alloy   8
    2.2   The mechanical attributes of the Ni–Ti alloy   13
       2.2.1   Superelasticity of Ni–Ti shape memory alloy   13
       2.2.2   The shape memory phenomenon of SMAs   21
    References   24

Chapter 3   The inferior vena cava filter   27
    3.1   Brief history of vena cava filters and their development   27
    3.2   The development of the SMA inferior vena cava filter   34
    3.3   Recent research into the SMA inferior vena cava filter   44
    3.4   Conclusion   62
    References   63

Chapter 4   The use of Ni–Ti SMA for intravascular applications   69
    4.1   Intravascular endoprosthesis   69
       4.1.1   Brief introduction to intravascular stents   69
       4.1.2   Applicaion of IVEPs   72
    4.2   Other intravascular devices   84
    4.3   Conclusion   88
    References   88

Chapter 5   Present and future orthopaedic applications   91
    5.1   Present orthopaedic applications of SMAs   91

	5.2	Future possible orthopaedic applications of SMAs		98
		5.2.1	The correction of scoliosis	98
		5.2.2	The union of fractures	99
		5.2.3	The use of orthotics	101
	5.3	Conclusion		104
	References			105

Chapter 6 Other applications of the SMA   **107**

    6.1 Dental arch wires   107
        6.1.1 A brief history of arch wires   107
        6.1.2 SMA thermal arch wires   107
    6.2 Other thermally activated SMA stents   111
        6.2.1 Introduction   111
        6.2.2 Tracheobronchial stents   111
        6.2.3 Urethral stents   114
    6.3 Ni–Ti clinical instruments   120
    6.4 Cranial aneurysm clips   128
    6.5 Intravascular grafts   132
    6.6 Conclusion   134
    References   134

Chapter 7 Conclusion   **141**

Glossary   **143**

Index   **151**

Foreword

Shape memory effect materials, particularly shape memory alloys (SMAs), are currently creating great interest in a number of scientific and commercial areas. The list of successful applications continues to grow and awareness amongst engineers of the attributes of these materials is no longer solely the domain of research laboratories.

This book demonstrates some of the uses of shape memory alloys in the medical field, giving both practical information and evidence of the efficacy of implant and non-invasive devices. Medical devices made from SMAs – both implants and instruments – are an ideal way of utilizing the unusual properties of these materials and the increased use of minimally invasive 'key-hole' surgical techniques will create new applications for SMAs. The rationale behind these developments is to be able to introduce a wire, or other suitably small cross-sectional form, into the body where its shape changes into a different, much larger, geometry when heat is applied. This shape-change phenomena can be arranged to do useful work within the body such as to constrict, dilate, push apart, or pull together body components. Thermal actuation of the devices may be from a variety of remote non-contact sources: from thermal energy contained within the body, direct electrical resistance heating, or infusion of a suitable media to trigger the memory effect.

A developing track record associated with *in-vivo* nickel–titanium SMA devices indicates excellent biocompatibility and corrosion resistance compared with the commonly available medical quality stainless and alloy steels. Indeed, the SMAs have secured an important place between titanium alloys and the stainless steels. Any problems which might occur with implanted devices being used in particular circumstances (such as nickel migration) have yet to be fully qualified, but are likely to be outweighed by the overall benefits, especially if viewed in the light of current techniques concerned with surface modification and coatings. A number of insulative or protective coatings can be applied if toxicity is considered to be a problem in specific applications.

Apart from the shape changes available from SMA, a number of applications have been realised from some of the generic forms of the materials. Guide wires made from super elastic nickel–titanium have found extensive use as catheter guidance systems because of the ability of the wires to assume the extreme deformation required to negotiate tortuous paths within the vascular system and their capability to return to their original shape when removed from the body. The super elastic form of the material does not require additional thermal input to maintain the original guide wire shape. Equally, a series of steerable guide wire bundles using the thermally acuated shape change are being incorporated into catheters where the shape memory effects are electrically induced in each wire to deflect the catheter tip and steer it to the required position. Another application is 'multi-stage' SMA where shape memory effects are available at discreet thermal increments. In this case there may be three or four of these increments, each stage demonstrating a particular shape change of the material.

Direct product manufacture using sintering techniques is also under development. Here the shape is induced at the primary manufacturing stage without the need for additional thermal processing to implant memory. This, therefore, allows very cost effective devices to be produced.

Global research associated with shape memory effect materials is extensive. Medical implant and instrument developments look particularly exciting and may eventually lead to new improved techniques demonstrating considerable clinical and economic benefits. This valuable book illustrates current state-of-the-art technology and indicates the potential benefits of this interesting and exciting technology.

Tony Anson
Brunel University, Middlesex, UK
1995

Preface

Since their discovery in the early 1960s, shape memory alloys have been developed for a wide range of applications. However, it was not until the 1970s that the use of memory metals and their possible applications in the medical field started to receive considerable attention. Since then a multitude of clinical functions have been looked into, all of which use the nickel–titanium (Ni–Ti) shape memory alloy which has been shown to have good biocompatibility properties.

Although a relatively new alloy, a few medical applications incorporating its unique properties have already been developed to the commercial stage:

- vena cava filters;
- dental arch wires;
- orthopaedic clips and staples.

All of these use the unique thermo-mechanical properties portrayed by the alloys in the shape memory family. Further development in different medical fields is also being carried out with increasing interest, these include:

- intravascular stenting
- tracheobronchial stenting
- ureteral stenting
- rectal stenting
- clinical tools
- aneurysm clips
  and further orthopaedic applications.

All of these areas have been investigated to some degree and all have shown, at least at an initial level, the shape memory alloy to be an improvement over current methods of treatment.

All in all the use of memory metals in medical applications has a bright future with new and ever more varied ideas being put forward for its possible uses.

Acknowledgements

With many heartfelt thanks to Miss J Watkins and Mrs D Hillman for their vast input of hours and patience into this project, and personal support from Miss T Ardani. Further thanks to Mr E Maylia, Dr T Flint, and Mr P Weaver at the University of Wales College of Cardiff (UWCC) for their help, criticism, and guidance in the extension of the original concept.

CHAPTER 1

Introduction to Shape Memory Alloys (SMAs)

1.1 The history of shape memory alloys

In the scientific literature there is much confusion surrounding the date when the shape memory effect was first reported.

It has been suggested that the effect was first referenced in 1938 by Greninger and Mooradian (**1**), others believe that Chang and Read (**2**) noted the phenomenon in 1951. In fact it was not until 1963 that the shape memory effect was first officially reported (**3**).

Buehler and his co-workers (**3**) at the US Naval Ordnance Laboratory, Silver Springs, Maryland, were first to report the shape memory effect. They discovered the property in an alloy of nickel (Ni) and titanium (Ti) containing equiatomic properties of each element. They named the new alloy nitinol (Ni–Ti–Nol, Nol = naval ordnance laboratory). The discovery of nitinol sparked off many investigations into the shape memory effect and attempts were made to understand the mechanics of its behaviour. From their investigations many alloy systems have been identified which show the shape memory phenomenon, such as Cu–Al, Ni–Al, Cu–Al–Ni, Cu–Zn, Sn and Cu–Zn–Al.

The use of Ni–Ti alloy for medical purposes was first reported in 1971 by Andreasen and Hilleman (**4**) and in 1976 Castleman *et al*. (**5**) reported that the biocompatibility of the alloy was sufficient for use as a biomaterial. Since then its possible applications in orthodontics and orthopaedics have been widely researched and a few applications have been developed commercially, these include dental arch wires, vena cava filters, and orthopaedic bone implants, all of which are in use in hospitals around the world.

Other on-going investigations into the possible use of the Ni–Ti alloy include arterial and tracheal stents and medical instrumentation.

1.2 The mechanics of the shape memory effect

Characteristics associated with shape memory metals are due to the unique interaction between the martensite and austenite crystal structures of shape memory alloys (SMA). This interaction between the crystal lattices was first reported by Buehler *et al*. They noted that the shape memory effect was related

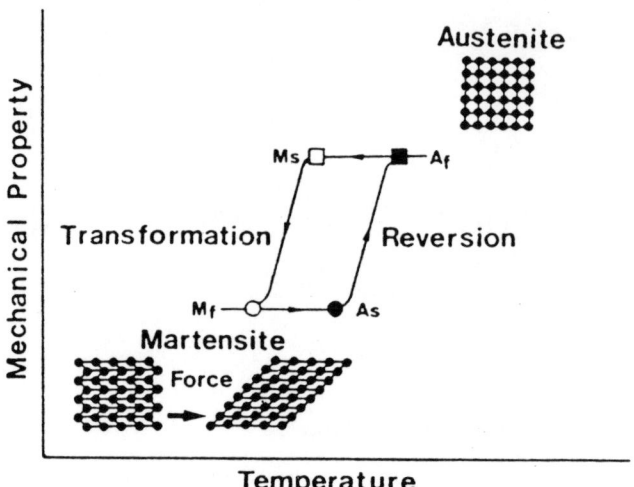

Fig 1.1 The crystal structure interaction as a result of temperature (6)
M_S Temperature of start of martensite transformation
M_f Temperature of finish of martensite transformation
A_S Temperature of start of reverse of austenite transformation
A_f Temperature of finish of reverse austenite transformation

to the inherent capability of the SMA to alter its atomic bonding as the function of heat.

A memorized shape is imprinted into the SMA by holding the SMA in the desired 'memory' shape at a high temperature. The desired shape must also be held during the cooling process. On cooling, the alloy passes through what is called the transition temperature range (TTR). Below this range the alloy can be released from its mould or constraints and will appear pliable and soft. The mechanics involved with the crystal structure interaction and TTR can best be explained with the aid of a diagram (Fig. 1.1).

On cooling from the setting temperature, a temperature M_s is reached. At this point the crystal structure starts to change from, what has up to now been completely austenite, to a martensite structure (a change from face centre cubic to hexagonal close packed). This martensite transformation is completed by the temperature M_f, where all the metal is consisting of the martensite phase.

On reheating, a temperature of A_s has to be reached, note $A_s > M_f$, before the metal's crystal structure starts to change back to its austenite phase. The transition temperature range is the temperature range between A_f, where all the crystal structure is made up of austenite structure, and M_f, which has been explained before. Below the M_f the allow may be deformed with no loss to its shape memory properties, so long as the change is kept below the 8 percent surface, plastic deformation limit.

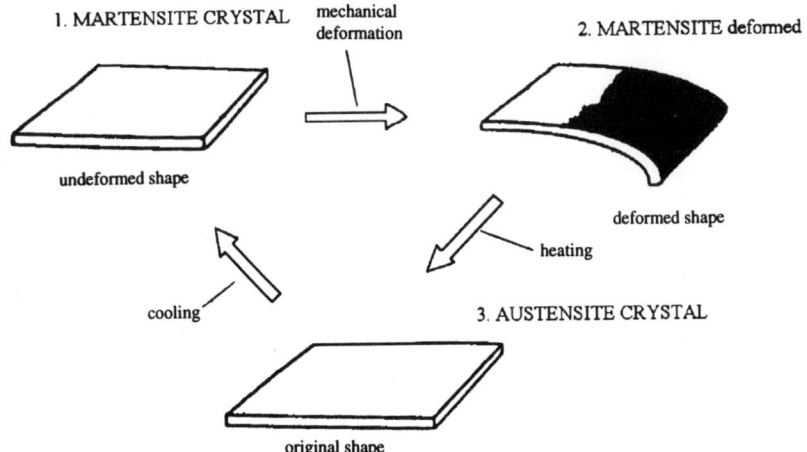

Fig 1.2   Schematic diagram showing the reversible martensite transformation

The martensite crystal structure is made up of a lattice of 'zig-zagged' (overlapping) atoms, which allows the metal to be soft and ductile. The austenite crystal structure, on the other hand, consists of an ordered row–column matrix of atoms, which means the metal is hard and rigid.

The shape memory effect occurs due to the fact that, as the temperature is raised above A_s, a stress forms in the crystal structure and the atoms return to their austenite phase. The shape that the crystal structure forms is the one that was 'memorized' into the austenite phase of the metal during the imprintation process. This transition to austenite from the martensite crystal structure is completed at A_f.

This is a reversible process and is shown diagrammatically in Fig. 1.2.

References

(1) GRENINGER, A. B., and MOORADIAN, V. G. *Trans Met Soc AIME*, 1938, **128**, 337.
(2) CHANG, I. C., and READ, T. A. *Trans Met Soc AIME*, 1951, **191**, 47.
(3) BUEHLER, W. J., GILFRICH, J. V., and WILEY, R. C. Effect of low-temperature phase changes on the mechanical properties of alloys near composition Ti–Ni, *J. Appl Phys*, 1963, **34**, 1475–1477.

(4) ANDREASEN, G. F., and HILLEMAN, T. B. An evaluation of 55 cobolt substituted nitinol wire for orthodontics, *J. Am Dent Assoc*, 1971, **82**, 1373–1375.
(5) CASTLEMAN, L. S., MOTZKIN, S. M., ALICANDRI, F. P., BONAWIT, V. L., and JOHNSON, A. A. Biocompatibility of nitinol alloy as an implant material. *J. Biomed Mater Res*, 1976, **10**, 695–731.
(6) TAKAMI, M., FUKUI, F., and SAITOU, S. *et al*. Application of a shape memory alloy to hand splinting, *Prosthetics & Orthotics Int*, 1992, **16** (1), 57–63.

CHAPTER 2

Characteristics of Shape Memory Alloys in Medical Applications

When a new material is applied to the medical field it must fulfil a number of requirements, covering the areas of mechanical, chemical, and biological reliability, before it will be classed as a biomaterial.

The Ni–Ti alloy is the only SMA that can pass all these criteria, and is, therefore, the only shape memory alloy in medical use. It has two main characteristics:

(a) good biocompatibility
(b) unique mechanical properties:
 – superelasticity;
 – shape memory effect.

2.1 Biocompatibility of Ni–Ti alloy

Due to the fact that the Ni–Ti alloy is a relatively new metal in the medical field, there have been few reports on its biocompatibility. Groups that have carried out research in this area include the following, Cutright *et al.* (**1**), Castleman *et al.* (**2**), Ohnishi *et al.* (**3**), Prince *et al.* (**4**), Putters *et al.* (**5**), Randin *et al.* (**6**), and Speck *et al.* (**7**). The first four investigators considered the effect of *in vivo* implantation, whilst Putters *et al.* carried out an *in vitro* cell culture test. Randin *et al.* and Speck *et al.* analysed the corrosion behaviour of Ni–Ti alloy.

2.1.1 In vivo *testing of Ni–Ti alloy*

Cutright *et al.* implanted lengths of Ni–Ti (50 at. percent Ti approximately) into the subcutaneous layers of skin of 45 rats. The wire was 15 mm long, 0.76 mm in diameter and was left *in situ* for up to 9 weeks. The tissue response was observed over a period of time ranging from 3 days to 9 weeks. The group used results gained from previous investigations to compare the Ni–Ti tissue

Table 2.1 Tissue response to implanted Ni–Ti wire in rates in tests performed by Cutright *et al*. (1)

Time elapsed	Comments on Ni–Ti implant
3 days	Healing at wound site was beginning. Odema was prominent. Loose fibrous connective tissue stroma had formed around implant.
1 week	Connective tissue stroma had become dense. Proportion of collagenous tissue in fibrous connective tissue had increased in proportion to observation of 3 days. Swelling around the implantation area has almost disappeared.
3 weeks	Dense fibrous connective tissue with large amounts of collagenous fibres had formed around Ni–Ti implant. Fibroblasts not growing compared to initial period.
4–9 weeks	Surrounding tissue grafts stabilized, fibrous connective tissue had become dense and only a small number of chemically inflamed cells remained.

Note: The results from previous stainless steel investigations showed slight differences, the healing process began one to two weeks after implantations, and dense fibrous tissue had formed after five to six weeks. However, after the nine week period, the results were more or less the same.

reaction with that of stainless steel. This allowed the Ni–Ti to be assessed relative to a known biomaterial. The results are shown in Table 2.1.

The paper concluded:

> When compared to stainless steel in previous experiments, the nitinol was indistinguishable from the stainless steel at similar time periods.
>
> On the basis of the results in this experiment, it appears that 55–nitinol histologically compares favourably with stainless and could be used on deep tissues.

Castleman *et al*. used a high purity nickel–titanium alloy to produce prototype bone plates, fastening screws and various medical instruments. The group attached the plates by means of the fastening screws to the unfractured femoral bone of 12 dogs. Cobalt–chromium alloy (Co–Cr alloy) bone plates and screws were used as a control in four dogs. Co–Cr is a widely utilized biomaterial and sham operations were performed on four beagles as a further control.

The animals were set into groups (see Table 2.2), and sacrificed after set periods.

Table 2.2 Exposure schedule of dogs involved in Castleman's *et al*.'s investigation (2)

Group number	Number of dogs		Duration of exposure to plates (months)
	With plates	Without plates	
I	4	1	3
II	4	1	6
III	4	1	12
IV	4	1	17

Table 2.3 Evaluation by neutron activation analysis of tissue samples taken from implanted dogs by Castleman et al. (2)

Tissue	Nitinol		Co–Cr alloy			
	Nickel		Cobalt		Chromium	
	No. of samples	Presence in tissue	No. of samples	Presence in tissue	No. of samples	Presence in tissue
Bone adjacent to implant	10	?	4	No	4	Yes
Muscle adjacent to implant	7	No	2	No	2	No
Liver	3	No	3	No	2	No
Spleen	6	No	2	No	2	No
Heart			1	No	1	No
Lung	4	No	1	No	1	No
Brain			2	No	2	No
Fat	5	No	2	No	2	No
Kidney	6	No	2	No	2	No

Gross chemical, radiological and morphological observations were made at the site of the implant. Further to that, histological examination and neutron activation analysis were performed on samples of muscle and bone taken adjacent to implantation site and from tissues removed from cerebrum, lung, spleen, liver and kidney. The results obtained are shown in Table 2.3. The plates and screws were removed after sacrifice and examined microscopically for signs of corrosion or other imperfections.

From the results, Castleman et al. were able to conclude the following:

(a) In all groups (I–IV) there were no signs of either generalized or of localized corrosion that could be attributed to the reaction between the metallic surfaces and adjacent tissues for either type of implant.
(b) Gross clinical and radiological observations indicated that there was no discernible difference between 'sham' operated dogs and implanted dogs from any of the four groups.
(c) In all implanted dogs a well defined fibrous capsule covered the implant plate.
(d) There was no significant histological difference between tissue and bone samples taken from the adjacent area to either implant and sham operated dogs. Also, no significant histological difference was found between the 'sham' operated dogs and implanted dogs with either alloy in tissue taken from liver, spleen, lung, kidney and brain.
(e) There was some evidence by neutron activity analysis that bone adjacent to a Co–Cr alloy implant was contaminated with chromium after exposure of 17 months.
(f) There was evidence by neutron activity analysis that no contamination occurred due to either implant in the liver, lungs and brain.

Table 2.4 Results gained by implantation of Ni–Ti plates in mature rabbits by Ohnishi *et al*. (3)

Time elapsed	Comments
2 weeks	Connective tissue had formed between bone and implant plate. Abnormally inflamed cells were not noticed.
4 weeks	Most of the connective tissue between plate and bone was changing due to the formation of ostesial tissue, which gradually covered the plate.
6 weeks	Ostesial tissue was noticeable even on surfaces where the bone and plate did not touch.

Note: Compared to the results taken from similar tests on stainless steel implants, it was shown that new bone development was slower and some pitting was noticed on the stainless steel plate, that did not occur on the Ni–Ti implant. Also ions of various metals (Fe, Cr, Ni) were observed in the stainless steel implant whereas Ni–Ti ions were hardly observed at all.

Castleman *et al.*'s report showed that the Ni–Ti alloy implant was at least as good, if not better, than the Co–Cr alloy implant, and held promise as a superior functional material for internal use.

Ohnishi *et al*. (3) implanted Ni–Ti plates into the surface of fibular in mature domestic rabbits, for a period of 6 weeks and reported on the reaction of the living tissue. The group also used stainless steel and reported on the difference between the two metals.

The results gained from the Ni–Ti alloy implanted are shown in Table 2.4.

Prince *et al*. implanted, *in vivo*, three designs of SMA inferior vena cava (IVC) filter. The group's aim was to demonstrate the non-thrombogenic nature of the SMA filter, and hence show its ability for use as a clinically applicable device. Prince *et al*. implanted twenty-seven filters with varying degrees of polishing into seventeen animals (canines n = 16, sheep n = 1), that were sacrificed at times between one week and four years.

The three designs: one-wire; two-wire; and seven-wire types (shown in Table 2.1) were all comparable in amount of wire used. The animals were checked on a weekly basis for signs of toxicity, and gross autopsy examinations were performed. The results are shown in Table 2.5 – no animal showed signs of illness during weekly inspection.

Prince *et al*. noted out that the results were comparable to stainless steel (and hence the Greenfield IVC filter) and concluded that:

> In addition to showing the filter to be biocompatible and sufficiently non-thrombogenic to avoid occluding the vena cava, these studies have shown the final seven-wire filter design to be mechanically suitable as well.

2.1.2 In vitro *testing of Ni–Ti alloy*

In a biocompatibility report in 1992, Putters *et al*. (5) compared the effects of SMA Ni–Ti with its constituent elements (nickel and titanium).

Table 2.5 Incidence of thrombosis for each filter studied by Prince et al. (4)

Animal #	Filter design Clean	Surface finish	Location	Angiography at 1 week	Autopsy	Gross path	Time to autopsy
1	N	Rough	IVC AR*	Patent	Patent	Patent	9 weeks
1	N	Rough	IVC BR	Partially occluded	Occluded	Occluded	5 weeks
2	N	Rough	IVC BR	Occluded	Occluded	Occluded	7 weeks
2	N	Rough	IVC BR	Partially occluded	Partially occluded	Partially occluded	6 weeks
3	N	Rough	IVC BR	Partially occluded	None		
4	N	Rough	IVC BR	Occluded	Occluded	Occluded	4 months
5	N	Rough	IVC BR	Occluded	Occluded	Occluded	9 weeks
6	N	Rough	IVC BR	Occluded	Occluded	Occluded	5 weeks
7	N	Rough	IVC BR	Occluded	Occluded	Occluded	3 weeks
8	Y	Rough	IVC BR	Patent	None		
9	Y	Sanded	IVC BR	Partially occluded	None		
10	Y	Sanded	IVC AR	Partially occluded	Partially occluded	Partially occluded	4 years
10	Y	Sanded	IVC BR	Partially occluded	Partially occluded	Partially occluded	11 weeks
11	Y	Sanded	IVC BR	Patent	Patent	Patent	11 weeks
12	Y	Electropolished	IVC BR	Partially occluded	Partially occluded	Partially occluded	4 weeks
13	Y	Electropolished	IVC AR	Patent	Patent	Patent	10 weeks
13	Y	Rough	IVC BR	Partially occluded	Partially occluded	Partially occluded	10 weeks
14	Y	Sanded	IVC BR	Patent	None		
15	Y	Rough	IVC BR	Partially occluded	Partially occluded	Partially occluded	6 months
15	Y	Sanded	SVC	Patent	Patent	Patent	6 months
15	Y	Electropolished	IVC AR	Patent	Patent	Patent	6 months
16	Y	Rough	IVC AR	Patent	Patent	Patent	1 week
16	Y	Sanded	IVC BR	Patent	Patent	Patent	1 week
16	Y	Electropolished	SVC	Patent	Patent	Patent	1 week
17	Y	Rough	SVC	Patent	Patent	Patent	1 week
17	Y	Sanded	IVC AR	Patent	Patent	Patent	1 week
17	Y	Electropolished	IVC BR	Partially occluded	Partially occluded	Partially occluded	1 week

Note: AR = above renal veins   BR = below renal veins   *Migrated to pulmonary artery trunk

Table 2.6 Typical chemical composition of the materials investigated by Randin et al. (6)

Materials	Ni	Fe	Cr	Co	Mo	Al	C	S	Mn	Si	Cu	Others	Remarks
Ni200	99.5	0.15					0.06	0.005	0.25	0.05	0.05		
CuNi 25	24–25	<0.2					<0.03	<0.02	<0.5		b		Coin alloy
$Ni_{50}Al_{50}$	68.5					31.5							Coloured intermetallic compounds
$Ni_{60}Al_{40}$	76.5					23.5							
$Ni_{70}Al_{30}$	83.6					16.4							
WC + Ni	7.0											WC b	Hard metals
$TiC + Mo_2C + Ni$	17											Mo_2C 12, TiC b	
White Gold	14.5										5.5	Au 75.0, Zn 5.0	Jewelry alloy
FN 42	42	b											Controlled expansion alloys
Nilo Alby K	29.5	b					0.02		0.80	0.15			
Ni–Ti	55											Ti 45	Shape-memory alloy
AISI 303	8–10	b	17–19		<0.6		0.05	>0.15	0.3				Austenitic stainless steels
12/12	12–14	b	12–14				<0.15	<0.03					
AISI 304	8–12	b	18–20		—		<0.08	>0.03					
AISI 316F	10–14	b	16–18		2–3		<0.08	<0.15					
316 PX	13.2	b	17.7		2–6		0.03	0.07					
AISI 316L	10–14	b	16–18		3–4		<0.03	<0.03					
Hastelloy X	b	18.5	22	1.5	9		0.10		0.5	0.5			Superalloy
Phydur	52	b	37			4	0.02		0.02	0.06			Precipitation hardening alloy
Coatings:													
NiP	b											$P \approx 8$	Electroless nickel
NiP/450°C	b											$P \approx 8$	
PdNi	b											PD 75–80	
SnNi	35											Sn 65	
Cr			100										
Sandvik	—	b	18										Ferritic stainless steels
Shomac	0.18	b	30				0.03	0.3					
Co				100									
$Co_{50}Al_{50}$				68.6		31.4							Coloured intermetallic compound
WC + Co	—		7									WCb, (TANb) C1	Hard metals
WC + Co	—		10									WCb, (TaNb) C1	
Stellite 20			45			2.5						W 18.0	
Ti		33										Ti 99.7	

Note: b refers to the balance of material made up of the indicated element

The group used an *in vitro* experimental set up and compared the effects of samples on the mitosis of cell cultured human fibroblasts. Mitosis is the process of cell division that enables the body to replace dead cells.

Putters *et al.* found that nickel showed significant detrimental effects on fibroblasts mitosis, whereas titanium and Ni–Ti showed no effect.

Putters *et al.* concluded:

> ... Nitinol is to be considered in this respect biocompatible and comparable to titanium, which would seem to justify application as a surgical implant ...

Randin *et al.* investigated the corrosion behaviour of many nickel alloys, including Ni–Ti in artificial sweat **(6)** (Table 2.6). Nickel and nickel salts can cause contact dermatitis if exposed to the skin. In two reports **(8) (9)** discussing the effects of nickel on skin, both commented on eczema caused by the element or its salts and both warned of the problems caused by nickel and advised against its use. Randin's experiment was designed to show the varying resistance to corrosion (and hence to nickel salts) of various nickel alloys.

Using an electro-chemical method to assess corrosion, Randin concluded that the Ni–Ti alloy was very resistant to pitting and hence the likelihood of contact dermatitis was low.

In 1980 Speck and Fraker **(7)** published results gained from *in vitro* corrosion tests performed on a group of alloys used in orthopaedics and orthodontics (Table 2.7). The metal samples were made into anodes with areas between 0.5 cm^2 to 0.8 cm^2, and these sample electrodes placed in various physiological solutions (Table 2.8). An electrical potential was applied and its value was raised from 0 volts to the breakdown potential of the given specimen.

Table 2.7 Alloys tested and their composition (wt percent) by Speck *et al.* (7)

Element	316 stainless steel	Cast Co–Cr alloy	Ti–Al–4V ELi	MP35N Co–Ni–Cr–Mo	Ni–Ti
Hydrogen	–	–	0.0058	–	–
Carbon	0.20	0.22	0.013	0.01	–
Oxygen	–	–	0.10	–	–
Nitrogen	–	0.05	0.011	–	–
Aluminium	–	–	6.2	–	–
Silicon	0.47	0.97	–	0.03	–
Phosphorus	0.021	0.006	–	0.008	–
Sulphur	0.002	0.006	–	0.004	–
Titanium	–	–	Remainder	0.74	44.94
Vanadium	–	–	4.05	–	–
Chromium	17.18	27.9	–	20.25	–
Manganese	1.81	0.48	–	0.02	–
Iron	Remainder	0.38	0.15	0.53	–
Cobalt	–	Remainder	–	33.64	1.60
Nickel	13.68	<0.05	–	34.7	53.46
Molybdenum	2.14	6.27	–	9.64	–
Copper	0.20	–	–	–	–

Table 2.8 Physiological solutions used by Speck et al. (7)

Solution	Composition (pH)
A	Hanks solution (7.4)
B	Hanks + cysteine* (1.5)
C	Hanks tryptophan† (7.4)
D	Solution B + HCl (2.2)

*Cysteine added to produce 0.1 M solution
†Tryptophan added to produce 0.002 M solution

The breakdown potential was the potential at which a relatively sharp increase in the ionization of the anode, and hence an increase in the current density, occurred. Each material was tested three times to ensure the reproducibility of the results and the magnitude before breakdown gave a direct indication of the corrosion resistance of the subject alloy. This 'anodic polarization' technique has been shown (10)–(12) to give a good indication of the corrosion resistance of implant materials.

Speck and Fraker used electrolytes based on Hanks (13) solution (Table 2.9), which has a pH of 7.4, and all tests were performed at 37°C. The pH of the electrolyte was altered during experimentation and cysteine and tryptophan were added to separate Hanks solutions to show the effects of amino acids. The results gained by Speck and Fraker led them to conclude that titanium and its alloys, NiTi SMA included, were more resistant than all the other clinically utilized alloys tested.

From the small number of investigations carried out it can be shown that the biocompatibility of the Ni–Ti alloy is at least equivalent to Co–Cr and stainless steel, both of which are widely used in medical fields.

Table 2.9 Hanks solution used by Speck et al. (7)

Solution A:	160 g NaCl 4 g $MgSO_4 \cdot 7H_2O$ 8 g KCl in 800 ml H_2O
Solution B:	2.8 g $CaCl_2$ in 100 ml H_2O
Solution C:	A + B + 100 ml H_2O + 2 ml $CHCl_3$ (chloroform)
Solution D:	1.2 g $Na_2HPO_4 \cdot 7H_2O$ 2.0 g $Na_2H_2PO_4 \cdot H_2O$ 20.0 g glucose in 800 ml of H_2O — — — → 2.0 ml $CHCl_3$ diluted to 1000 ml
Solution E:	1.4% $NaHCO_3$ = 7 g $NaHCO_3$ in 500 ml H_2O
Final solution:	50 ml C 50 ml D 24 ml E 900 ml H_2O few drops of chloroform

2.2 The mechanical attributes of the Ni–Ti alloy

Ni–Ti alloy shows two mechanical attributes that are widely exploited in its use. They are:

(a) superelasticity – stenting, orthodontics;
(b) shape memory phenomenon.

2.2.1 Superelasticity of Ni–Ti shape memory alloy

In the area of stenting the superelasticity of Ni–Ti has been widely overshadowed by the thermal SMA properties of the alloy. However, in the areas of esophageal and biliary stenting the Ni–Ti alloy has recently been investigated.

In 1993 Cwikiel *et al.* (**14**) discussed the utilization of self-expanding (superelastic) nitinol stents for the recanalization of esophageal lumen in patients with malignant esophageal strictures.

Malignant esophageal strictures are most often caused by some form of malignant cancer which commonly produces symptomatic dysphagia within the subject. Cwikiel *et al.* implanted forty superelastic nitinol 'Strecker' stents (Fig. 2.1) into patients between 21–92 years old (mean 70.4 years).

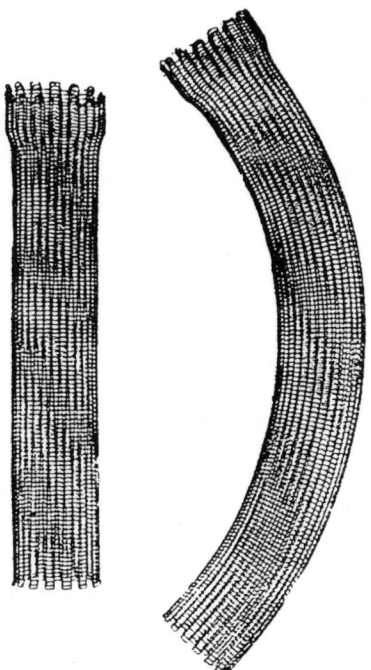

Fig 2.1 Graphical representation of the Cwikiel *et al.* 'Strecker' stent (14). (Courtesy of RSnA.)

The group's stent consisted of a metal mesh with internal diameter 18 cm and a length of either 10 or 15 cm. The prosthesis was implanted, and the immediate effect on the dysphagia of the patient was recorded (Fig. 2.2).

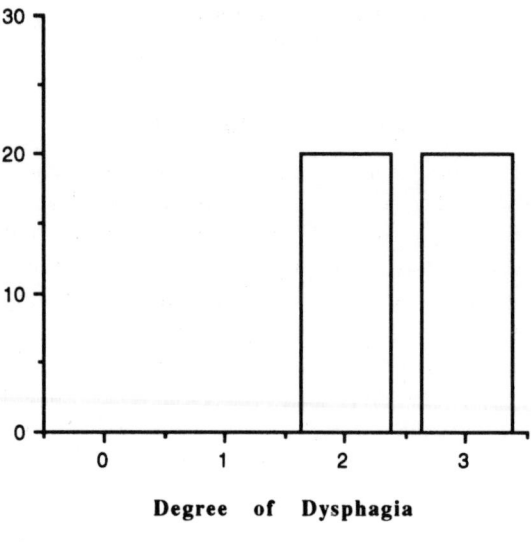

(a)

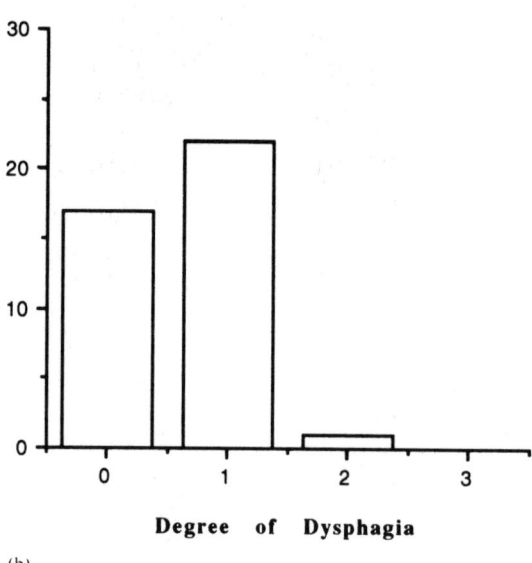

(b)

Fig 2.2 Immediate effect of stent insertion on dysphagia as observed by Cwikiel et al. (14): (a) before insertion; (b) after insertion. (Courtesy of RSnA.)

Cwikiel *et al.* summarized these studies by saying:

> The esophageal nitinol 'Strecker' stent appears to be valuable in the treatment of dysphagia in patients with malignant esophageal structures.
>
> Easy insertion, sufficient length, good longitudinal flexibility, and absence of shape element that might penetrate the esophageal wall are some of the advantages of this stent over other stents.

and the group concluded that:

> In our experience these stents are safe and effective in patients with dysphagia due to malignant esophageal structures.

In another paper published in 1993 a second group headed by W. Cwikiel (**15**) investigated the application of NiTi 'Strecker' stents in the management of benign esophageal strictures. The group's endoprosthetic material, size and placement were all comparable to that of Cwikiel *et al.*'s previous report.

The group perorally implanted ten stents into healthy Swedish land pigs weighing 20 kg and a further five endo-prostheses into human patients (Table 2.10).

Although Cwikiel *et al.* found minor signs of granulation tissue, ulceration, ischemia and fibrosis, the group noted that the dysphagia originally noted was greatly relieved. The group concluded that:

> There were no serious stent related complications.
>
> Presently, stent insertion in patients with benign esophageal strictures may be recommended as a temporary treatment in order to facilitate nutrition prior to esophageal resection. In selected elderly patients, it may be considered a permanent treatment.

In 1994 two groups: Bezzi *et al.* (**16**), and Rossi *et al.* (**17**), published papers referring to the same group of 35 superelastic (self-expanding) nitinol biliary stents which had been introduced to 19 patients suffering from malignant biliary obstruction.

Table 2.10 Findings from Cwikiel *et al.*'s five patients with benign esophageal strictures (15)

Patient/ age (y)/ sex	Stricture				Stent		
	Cause	Length (cm)	Minimum diameter (mm)	Location	Length (cm)	Expansion time	Observation period (months)
1/70/F	Caustic	10	4	Middle	15	2 w	16
2/65/M	Caustic	3	5	Middle	10	1 w	13
3/81/F	Peptic	8	5	Distal	15	1 w	4
4/82/M	Peptic	2.5	2	Distal	10	4 d	8
5/13/F	Caustic	14	1	Proximal middle	15	4 d	8

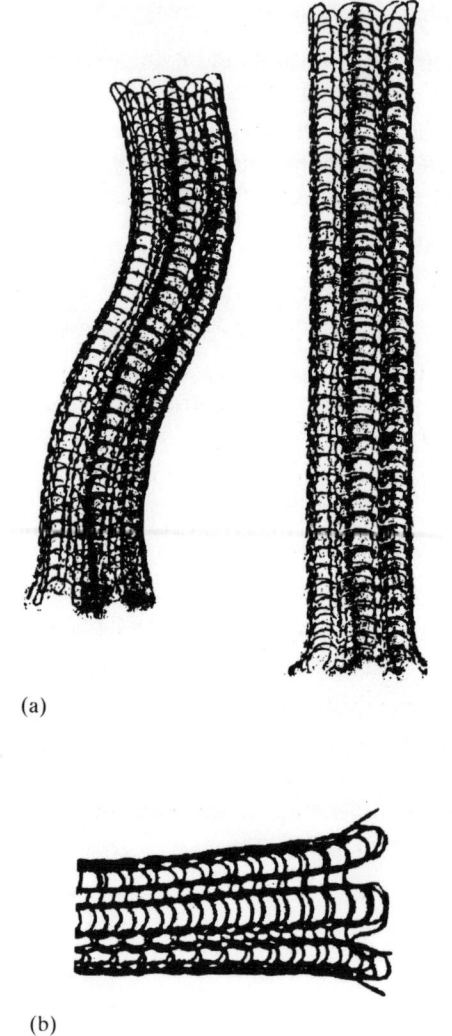

Fig 2.3 The nitinol 'Strecker' stent used by Bezzi *et al.* (16)

The nitinol stent had an internal diameter of 10 mm and a length of 6 cm (Fig. 2.3) and was based on the original Strecker stent (**18**). In most cases the stricture was balloon dilated before placement of the stent at the stenotic site occurred.

Delivery of the stent was achieved by constraining the stent under a plastic sheath, over a modified 10F catheter. Once at the stricture site the plastic sheath was removed and the stent expanded.

Bezzi *et al.* concluded that although

> The main limitation of the present study is the small number of patients.

the nitinol structure self-expanding stent was

> Adequate to re-establish bile flow in the occluded biliary tree. In addition they are easy and safe to use and caused no short term and few long term complications.

Rossi *et al.* (**17**) compared these results to other types of intra-biliary tree stents: Wallstents, Gianturco–Rosch Z stents, and Tantalum Strecker stents, and went on to discuss the place for all metal stents within modern management of biliary structures. Rossi *et al.* concluded that:

> Metallic stents represent a useful addition to the armamentarium of the radiologist undertaking biliary drainage for paliation of malignant destructive jaundice.
>
> The study indicates that, among the different stents used, the Wallstents and the nitinol Strecker stents are the most effective in providing long term paliation with reduced prevalence of late complications.

In orthodontics, specialized metal wires are used to realign teeth. This application uses the wires' elastic attributes to pull teeth that were misaligned back into the 'ideal arch' (see Fig. 2.4) and hence are called arch wires.

In 1971 Andreasen and Hilleman (**19**) reported the first test of Ni–Ti alloy as a replacement for conventional arch wire. They tested the Ni–Ti alloy against

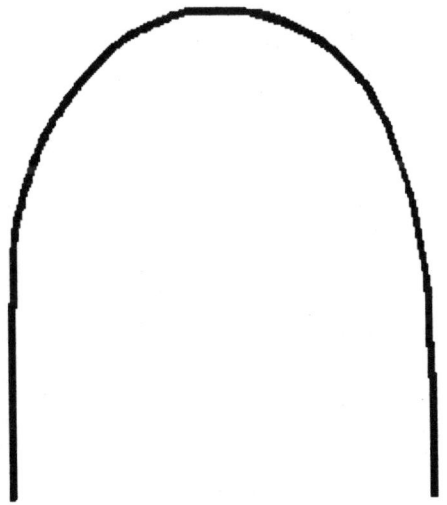

Fig 2.4 Representation of an 'ideal arch' (20)

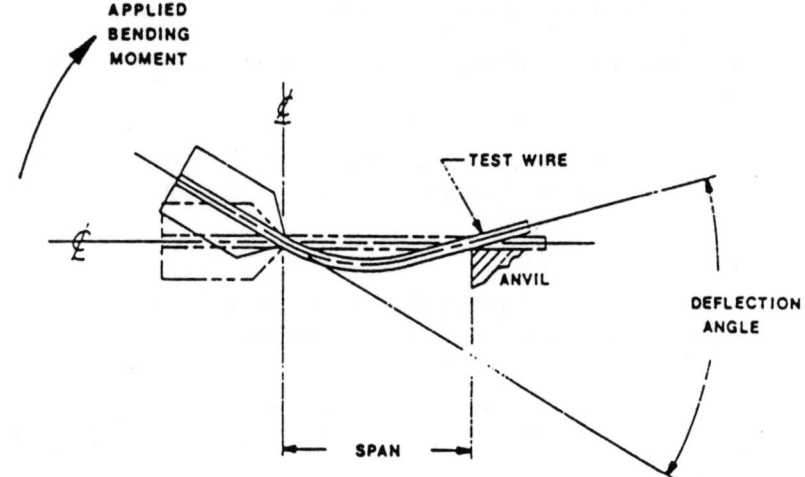

Fig 2.5 Bend test set up by Andreasen and Morrow (21) (lab and clinical analysis of nitinol wire)

other already medically applied arch wires, stainless steel and triple strand twistflex. The apparatus they set up enabled the middle tooth from a set of three to be offset by 1, 2 and 3 mm from the occlusion plane. The set up took measurements of the stiffness of the arch wire connecting the three teeth and the results were compared.

Andreasen and Hilleman concluded that Ni–Ti wire should be used in orthodontics as, in its unannealed form, it gave the best arch wire range of activation without plastic deformation.

In 1978 Andreasen and Morrow (**21**) reported on a more detailed investigation that looked into the effect of bending and torsion tests on Ni–Ti wire (Fig. 2.5), in comparison to stainless steel wire. They used both rectangular and round wire (Table 2.11) and discussed the results in the form of stored energy and spring rate (Figs 2.6–2.8).

The applied bending movement was increased up to a maximum deflection angle of 90°, where it was released and the resulting plastic deformation measured. In the test the Ni–Ti wires showed a resultant deformation of only 3–5 percent, compared to the 40–60 percent plastics deformation that occurred after the load was released in the stainless steel wires.

In the torsion test, one end of a length of wire was clamped, while a rotating jaw gripped the other end. Each samples was rotated through 720°, at which point the force on the rotating jaw was released and it was allowed to return to its natural point. The permanent set angle was measured.

The results of this test showed that the permanent rotational deformation of the Ni–Ti wire was at least five times less than that of the best stainless steel wire, 45° compared to 220°, respectively.

Table 2.11 List of the wires used in the testing by Andreasen and Morrow (20)

Material	Dimensions (inches)	Type
Stainless steel	0.020 dia	Round
Stainless steel	0.018 dia	Round
Stainless steel	0.016 dia	Round
Stainless steel	0.014 dia	Round
Stainless steel	0.012 dia	Round
Ni–Ti alloy	0.018 dia	Round
Stainless steel	0.017 × 0.025	Rectangular
Stainless steel	0.016 × 0.022	Rectangular
Stainless steel	0.016 × 0.016	Square
Ni–Ti alloy	0.019 × 0.025	Rectangular
Ni–Ti alloy	0.017 × 0.025	Rectangular

After completing both tests, the report went on to analyse the results by comparing the two different types of wire in the areas of 'stored' energy and spring rate:

(a) 'Stored' energy: this is the energy retained in the wire to enable it to return to its original formation. If we take two similar wires (0.017" × 0.025") made of stainless steel and the Ni–Ti alloy, then by comparing their bend characteristics, we can find the stored energy which is graphically described as the area under the force/bend angle graph (Fig. 2.6).

From comparing the two wires, where the maximum bend angle was 90°, it is obvious that the stainless steel wire has less 'stored' energy than that of the Ni–Ti wire (the area of the triangle is smaller). However, a 90° offset is

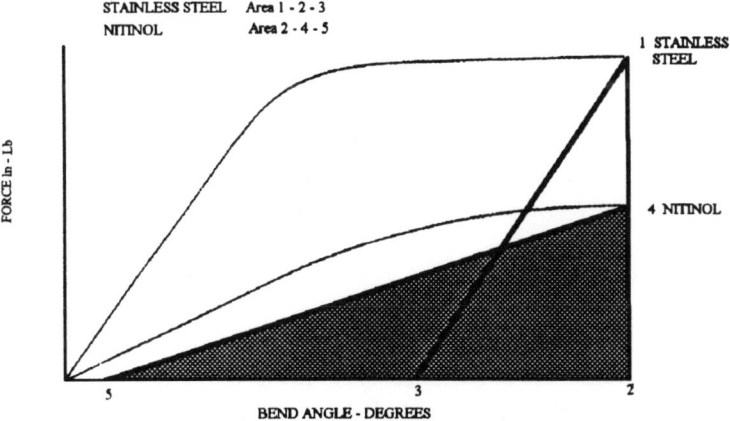

Fig 2.6 Comparison of the stored energy in nitinol and stainless steel (21), spring energy defined as the magnitude of the areas 1, 2, 3 and 2, 4, 5 for stainless steel and nitinol, respectively

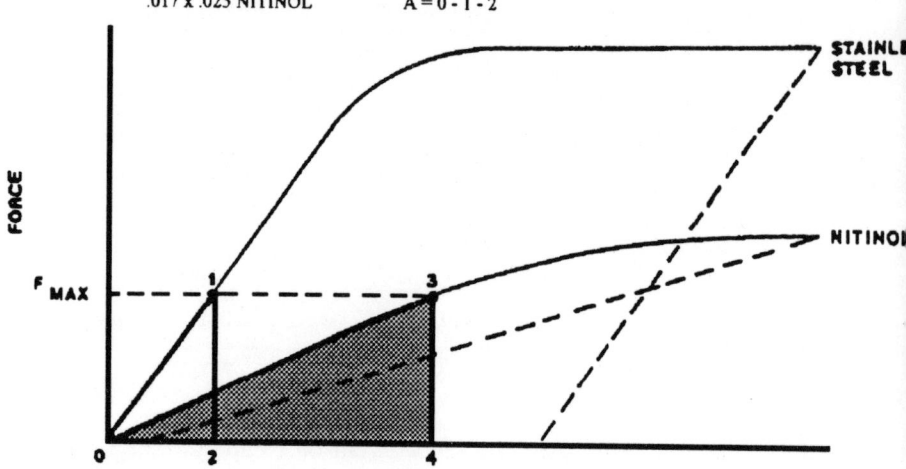

Fig 2.7 Comparison of the stored energy in nitinol and stainless steel wire when the force is limited (21), **spring energy defined as the magnitude of the areas 0, 1, 2 and 0, 3, 4 for stainless steel and nitinol, respectively**

unlikely to occur in a clinical situation, so a lower angular deflection was chosen (Fig. 2.7).

It is clear from these diagrams also that, for the same force, Ni–Ti alloy has more stored energy (area 4, 5, 2 compared to 1, 3, 2), and hence has a larger amount of work possible to move teeth.

(b) Spring rate: in the stored energy analysis, we considered that there was only one limiting factor affecting the implementation of the wire; this being the maximum force applicable for patients to remain comfortable.

However, this is not necessarily the case. In some situations the set deflection limits the use of the wires. Spring rate is an analysis using deflection based quantities. Consider the same example as for the 'stored' energy section (Figs 2.6 to 2.7), but instead of using a set force, a set deflection x is now substituted. Then a change in force dF must occur between the upper and lower limits of x (see Fig. 2.8).

dF is directly related to spring rate using the following equation

$$\text{Spring rate} = \frac{dF}{x} \qquad (1)$$

It is obvious by visually comparing the two plates that dF_s (the rate of change of force for stainless steel) is a lot larger than dF_n (the rate of change of force for Ni–Ti alloy), that is the slope of the stainless steel graph is steeper than that of the Ni–Ti and from calculations performed using equation (1), it can be shown that the spring rate of stainless steel is

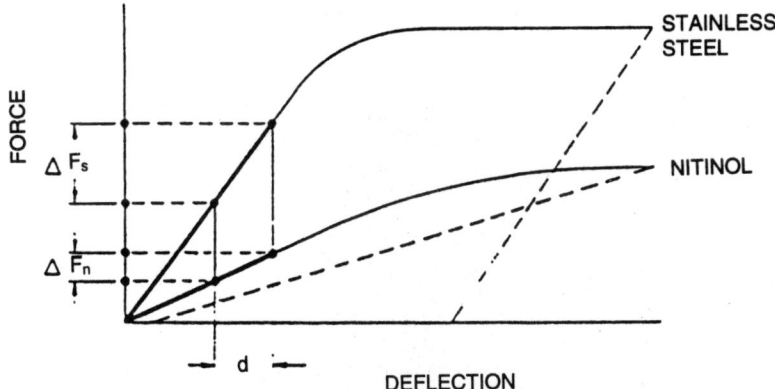

Fig 2.8 Comparison of the change in force between nitinol and stainless steel when undergoing a constant change in deflection (21)

approximately twice that of the Ni–Ti alloy. These results indicate that the Ni–Ti alloy would produce a lower, more constant, continuous force on the tooth or teeth than an equivalent stainless steel wire.

Andreasen and Morrow concluded that the Ni–Ti alloy could provide a larger step forward in orthodontic treatment, as it showed significant improvements over existing conventional wires.

Nitinol started its clinical application in 1972. Factor such as 'stored' energy, spring rate, and superelasticity make nitinol an ideal alloy for use in orthodontic arch wires.

2.2.2 The shape memory phenomenon of SMAs

In 1968 Buehler and Wang (**22**) produced a report discussing the Ni–Ti alloy constitution and the shape memory effect.

The group discussed two groups of Ni–Ti atomic ratios. The first was termed '55-nitinol' and consisted of stoichiometric quantities of both elements. The second group was '60-nitinol' and was a relatively nickel rich alloy.

The former of the two groups was the only one to demonstrate the shape memory effect. Buehler *et al.* discussed the mechanics of the 55-nitinol memory phenomenon and commented on the effects of constitution (Ni–Ti ratio) on transition temperature range and thus A_s. The discoveries of A_s variation around stoichiometric Ni–Ti as found by Buehler and Wang are shown in Fig. 2.9.

In a similar observation into alloying ratios published in 1976, Eckelmeyer *et al.* (**23**) arrived at very similar conclusions:

(a) That small increases in percentage titanium around stoichiometry

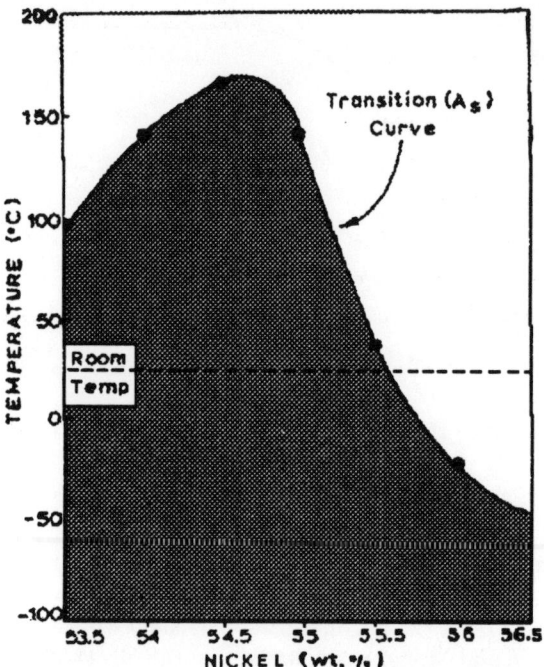

Fig 2.9 Martensite transition temperature (A_s) change with minor variations in the Ni–Ti ratio as observed by Buehler and Wang (22). (Courtesy of Elsevier Science Ltd, UK.)

increased the TTR substantially until a point is reached where no further increase affects the recovery temperature.
(b) Increases in nickel around stoichiometry causes substantial decrease in TTR.
(c) Large movements away from stoichiometry have a detrimental effect on the ability of the alloy to display the shape memory phenomenon.

In 1969 Buehler and Cross (**24**) also discussed the '55-nitinol' alloy. The group commented on the ability of lowering the TTR by a substitution of small quantities of cobalt for nickel. Buehler and Cross discussed the methods of alloying, the wire fabrication technique (Fig. 2.10), the unusual design

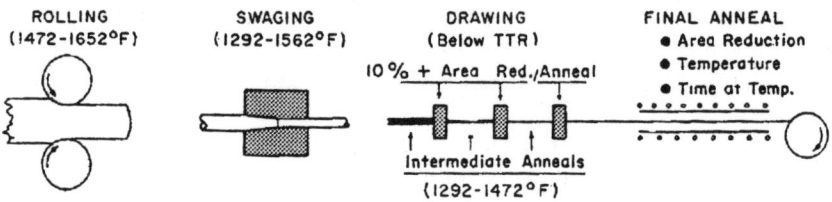

Fig 2.10 Diagram showing steps involved in converting a cast nitinol ingot into annealed wire as performed by Buehler and Cross (24). (Courtesy of Wire Journal International.)

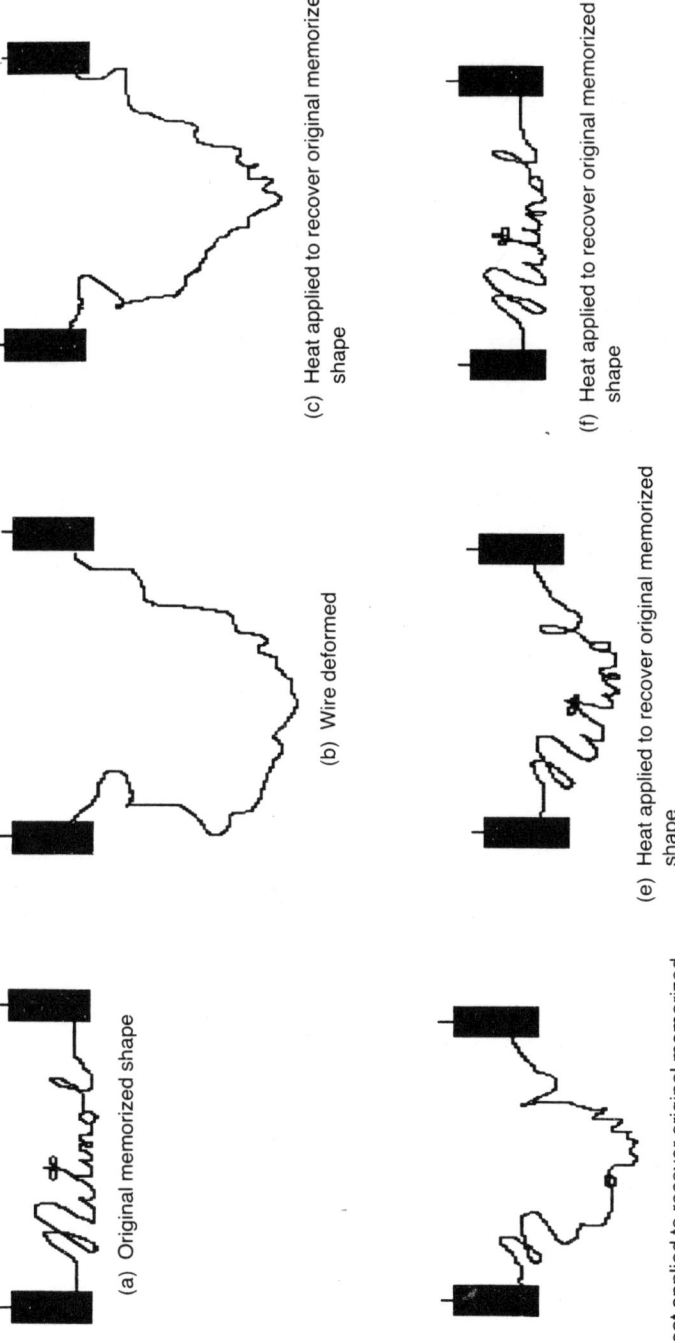

Fig 2.11 The recovery of memorized wire after heating electrically into the word nitinol – an example of the shape memory effect (22). (Courtesy of Wire Journal International.)

properties and concluded with a small section on application which Buehler and Wang categorized into four areas:

(a) self-erectable space structures
(b) thermally activated devices
(c) self-actuating fasteners
(d) principal component of heat mechanical energy converters.

None of these applications can easily be associated with medical utilization. This is probably because Buehler and Cross aimed to produce a metal for industrial application, and thus the oversight of a possible non-industrial use would not have been considered.

An Ni–Ti alloy of around equiatomic percentages of both elements exhibit the shape memory effect. However, slight changes in the atomic ratio or addition of small qualities of other elements can produce large changes in the alloy's transition temperature range. This ability to alter the alloy, and hence control the reaction of the shape memory alloy, has increased its potential in medical fields. Choosing an alloy whose austensite phase finishing temperature A_f is just below body temperature, 37°, allows an implant to remember and hold, in its rigid austensite state, the memory arrangement once introduced into the body.

Applications for the shape memory effect using Ni–Ti alloy were first discussed in 1975 by Civjan *et al.* (**25**). They set up a variety of experiments, ranging from the production of Ni–Ti tools such as clasps, chisels and tripod, and so on, to using Ni–Ti wire to realign teeth. (Note this was using its shape memory phenomenon to realign, not the aforementioned superelasticity.) The results gained from the experiment showed promise and allowed new application ideas to grow.

One of the main characteristics exhibited by SMAs has already been utilized to great effect. This is the ability of large complicated 'memorized' structures to be drawn in to small diameters (once under M_f), allowing percutaneous implantation and avoiding the risks of surgery. On placement the alloy retains its original shape (see Fig. 2.11).

The uses of the shape memory effect and alloys that possess it are discussed in more detail in subsequent chapters, as each case arises.

References

(**1**) CUTRIGHT, D. E., BHASKAR, S. N., PEREZ, B., JOHNSON, R. M., and COWAN, G. S. M. Tissue reaction to nitinol wire alloy, *J. Oral Surg*, 1973, **35**, 578–584.

(2) CASTLEMAN, L. S., MOTZKIN, S. M., ALICANDRI, F. P., BONAWIT, V. L., and JOHNSON, A. A. Biocompatibility of nitinol alloy as an implant material, *J. Biomed Mater Res*, 1976, **10**, 695–731.
(3) OHNISHI, HAMAGNCHI, NABESHIINA, MIYAGI, TUSJI, HAMADA, SUZKI, and SKIDA. *Proc. of 3rd Conf on Japanese Soc of Biomat*, 1982, **121** (in Japanese).
(4) PRINCE, M. R., SALZMAN, E. W., and SCHOEN, F. J. *et al*. Local intravascular effects of the nitinol wire blood clot filter, *Invest Radiol*, 1988, **23**, 294–300.
(5) PUTTERS, J. L., KAVLESAR SUKUL, D. M., and DE ZEEUW, G. R. *et al*. Comparative cell culture effects of shape memory metal (nitinol), nickel and titanium: a biocompatibility estimation, *Euro Surg Research*, 1992, **24** (6), 378–382.
(6) RANDIN, J. P. Corrosion behaviour of nickel containing alloys in artificial sweat, *J. Biomed Mater Res*, 1988, **22**, 649–666.
(7) SPECK, K., and FRAKER, A. Anodic polarization behaviour of Ti–Ni and Ti–6Al–4V in simulated physiological solutions, *J. Dental Research*, 1980, **59** (10), 1590–1595.
(8) WELLS, G. C. Effects of nickel on skin, *British J. of Dermatology*, 1956, **68**, 237–242.
(9) BRANDRUP, F., and LARSEN, F. Nickel dermatitis provoked by buttons in blue jeans, *Contact Dermatitis*, 1979, **5**, 148–150.
(10) HOAR, T. P., and WEARS, D. C. Corrosion-resistant alloys in chloride solutions: materials for surgical implants, *Proc. Roy Soc Sev*, 1966, **A294** (1439), 486–510.
(11) REVIE, R. W., and GREENE, N. D. Comparison of the *in vivo* and *in vitro* corrosion of 18-8 stainless steel and titanium. *J. Biomed Mat Res*, 1966, **3**, 465–470.
(12) FRAKER, A. C., BUFF, A. W., and YEAGER, M. P. Corrosion of titanium in physiological solutions, *Titanium Science & Tech*, 1973 (R. I. Jaffee and H. M. Burte Eds), (New York), Plenum Press, **4**, 2447–2457.
(13) HANKS, J. H., and WALLACE, R. E. Relation of oxygen and temperature in preservation of tissues by refrigeration, *Proc. Soc Exper Biol*, **71**, 196–200.
(14) CWIKIEL, W., STRIDBECK, H., and TRANBERG, K-G, *et al*. Malignant esophageal strictures: treatment with self-expanding nitinol stent, *Radiology*, 1993, **187**, 661–665.
(15) CWIKIEL, W., WILLEN, R., and STRIDBECK, H. *et al*. Self-expanding stent in the treatment of benign esophageal strictures: experimental study in pigs and presentation of clinical cases, *Radiology*, 1993, **187**, 667–671.
(16) BEZZI, M., ORSI, F., and SALVITORI, F. M. *et al*. Self-expandable nitinol stent for the management of biliary obstruction: long term clinical results, *JVIR*, 1994, **5**, 287–293.

(17) ROSSI, P., BEZZI, M., and ROSSI, M. *et al.* Metallic stents in malignant biliary obstruction: results of a multicenter European study of 240 patients, *JVIR*, 1994, **5**, 279–285.
(18) STRECKER, E. P., BERG, G., and SCHNEIDER, B. *et al.* A new vascular balloon-expandable prosthesis: experimental studies and first clinical results, *J. Intervent Radiol*, 1988, **3**, 59–62.
(19) ANDREASEN, G. F., and HILLEMAN, T. B. An evaluation of 55 cobalt substituted nitinol wire orthodontics, *J. Am Dent Assoc*, 1971, **82**, 1373–1375.
(20) ANDREASEN, G. F., and AMBORN, R. M. Aligning, levelling and torque control – a pilot study, *Angle Orthod*, 1989, **59** (1), 51–60.
(21) ANDREASEN, G. F., and MORROW, R. E. Laboratory and clinical analysis of nitinol wire, *Am J. Orthod*, 1978, **73**, 142–151.
(22) BUEHLER, W. J., and WANG, F. E. A summary of recent research on the nitinol alloys and their potential application in ocean engineering, *Ocean Eng*, 1968, **1**, 105–120.
(23) ECKELMEYER, K. H. The effect of alloying on the shape memory phenomenon in nitinol, *Scripta Metallurgica*, 1976, **10**, 667–672.
(24) BUEHLER, W. J., and CROSS, W. B. 55-Nitinol – unique wire with a memory, *Wire Journal*, 1969, **2**, 41–49.
(25) CIVJAN, S., HUGET, E. F., and DeSIMON, L. B. Potential applications of certain nickel–titanium (nitinol) alloys, *J. Dent Res*, 1975, **54** (1), 89–96.

CHAPTER 3

The Inferior Vena Cava Filter

3.1 Brief history of vena cava filters and their development

After certain surgical techniques, set diseases, or various injuries, clots may form within the body. These emboli may travel through the venous system and lodge within the pulmonary or cerebral arteries causing complications or death.

The prevention of pulmonary embolisms is of vital importance and can be achieved by various methods. In most circumstances a drug regime will be initiated to dispense or inhibit thrombosis. However, in a small proportion of cases, this method may be ineffective or dangerous to the patient's health and in such cases another technique must be implemented.

The closure of the inferior vena cava by external means such as ties, clips or sutures, is a surgical method of obtaining this desired result and was originally postulated by Homans (1) in the mid 1940s. Homans was the first to formulate a true basis for surgical interruption, and commented that interruption of the inferior vena cava was a rational method of preventing recurrent pulmonary embolisms. Homans also discussed the possible eventualities arising from venous thrombi formed below the popliteal region and commented on the risk of pulmonary embolism. He concluded that interruption of the embolism pathway at either the common femoral (Fig. 3.1) or common iliac arteries (Fig. 3.2) cut down the possibility of fatality by a considerable amount. Homans further commented briefly on the observations made after vena cava interruption but made no significant conclusions.

Pulmonary embolisms still remain an important cause of mortality in hospitalized patients, with approximately 200 000 deaths in the United States each year (2). Whilst most patients can be treated with anti-coagulant or fibrolytic therapy, there are some cases where this is constrained or fails and mechanical interruptions must be made. Up to the 1970s these interruptions could only be achieved by specialized sutures, staples or external vena cava clips (3)–(6). These interruptions forced the venous blood to return to the right-hand side of the heart by means of the many collateral venous pathways, which are too small to carry dangerous emboli. In 1970, Ochsner *et al.* (7)

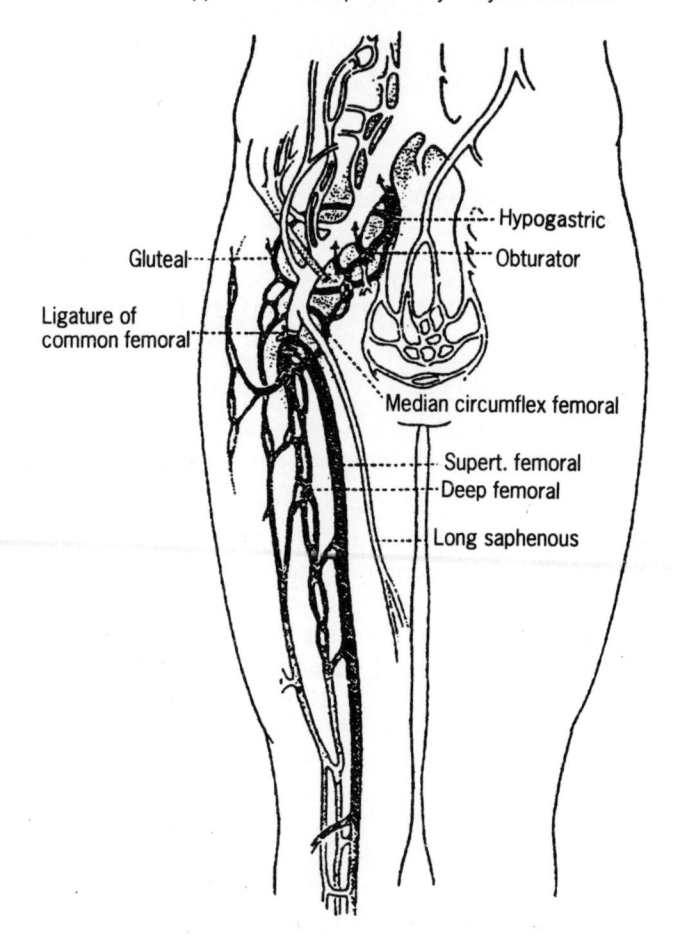

Fig 3.1 A diagrammatic representation showing the nature of venous return from the right leg and the collateral circulation when the common femoral vein is interrupted (1). (Courtesy of *Surgery, Gynecology and Obstetrics*, now known as the *Journal of the American College of Surgeons*.)

reported on the use of vena cava (caval) ligation to prevent such emboli. Ochsner *et al.* noted the rising cases of pulmonary embolism in hospitals around the USA, and discussed the problem of post operative sequelae, concluding:

> that caval interruption should be done only as a lifesaving procedure.

In a later publication, Blumenburg and Gelfand (**8**) commented that ligation was:

> an effective surgical measure to prevent pulmonary embolization arising from veins of the lower extremities.

However, all these interruption techniques still required surgery, even

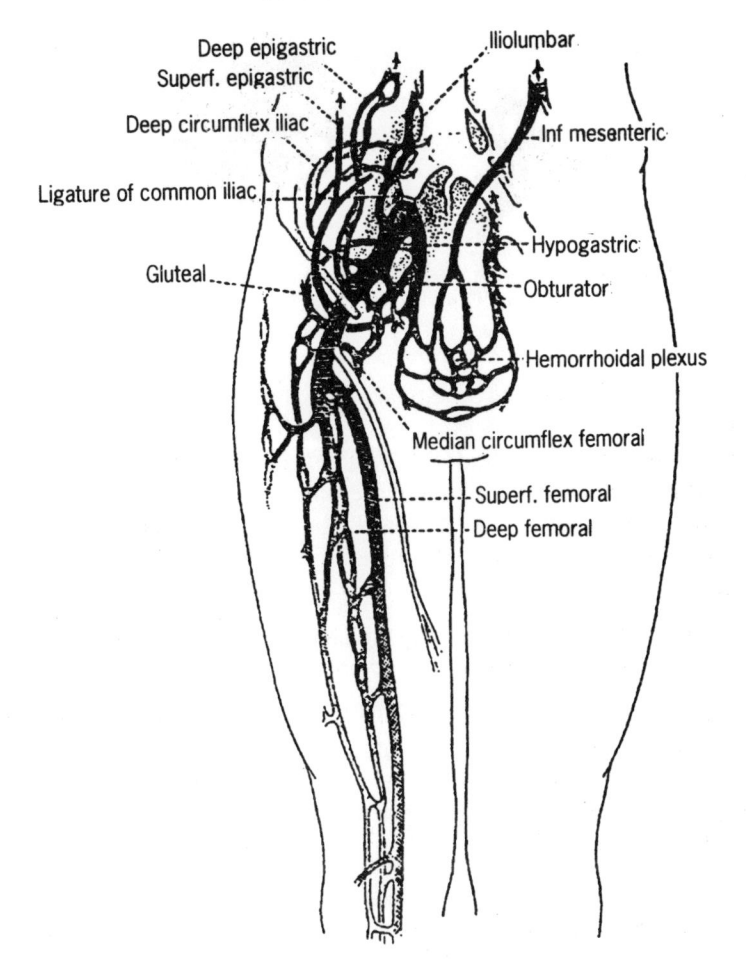

Fig 3.2 A diagrammatic representation (as in Fig. 3.1) showing the more abundant collateral venous pathway when the common iliac vein is interrupted (1). (Courtesy of *Surgery, Gynecology and Obstetrics*, now known as the *Journal of the American College of Surgeons*.)

though the patient may already by seriously ill from previous pulmonary embolism or underlying phlebothrombosis. To bypass this major drawback, many investigations were set up to look into a device that would cause less trauma, by gaining access to the vena cava through a peripheral part of the venous system. The first widely used device for this purpose was the Mobin-Uddin (MU) (**9**) (**10**) filter (shown in Fig. 3.3).

In a report produced in 1972, Mobin-Uddin *et al.* (**10**) published the experience of 100 patients who had the Mobin-Uddin transvenous filter placed within the inferior vena cava. The study group comprised of sixty-eight men and thirty-two women, aged between 27 and 85 years.

Fig 3.3 The Mobin-Uddin filter (10). (Reprinted by kind permission of K. Mobin-Uddin, G. M. Callard, H. Boboki, *et al*. Copyright (1972), Massachusetts Medical Society.)

The predisposing factor to pulmonary embolism in all 100 patients is shown in Table 3.1.

Of the group, thirty-five had died by the time of publication, with sixteen deaths within the first thirty days and the other nineteen over the further two years of the study. Ten cases of phlebitis occurred, though only mild lower extremity oedema was noted throughout the group. Mobin-Uddin *et al*. concluded:

> Transvenous caval interruption by the umbrella filter thus seems a safe and effective procedure for prevention of pulmonary embolism.

Although further studies also backed the idea that the MU filter was preferable over ligation or other surgical methods, it still suffered from unwanted occlusion of its gaps and increased flow turbulence within the lumen. Turbulence and its effects on thrombus formation was reported on by Stein and Sabbah (11) in 1974. They took eight dogs and placed two arteriovenous shunts

Table 3.1 Clinical data on patients with pulmonary emboli treated with the MU filter by Mobin-Uddin *et al*. (10)

Predisposing factor	No. of patients	Deaths		Recurrent emboli		Fatal emboli	
		Early	Late	Early	Late	Early	Late
Cardiac disease	43	7	11	0	2	0	0
Recent surgery	19	3	2	0	0	0	0
Nonspecific thrombophlebitis	14	0	1	0	0	0	0
Chronic lung disease	6	2	0	0	0	0	0
Cancer	4	1	3	0	0	0	0
Extreme obesity	3	1	1	0	0	1*	1*†
Trauma	3	1	0	0	0	0	0
Post partum state	1	0	0	0	0	0	0
Sickle cell disease	1	0	1	0	0	0	0
Undetermined	6	1	0	0	0	0	0
Totals	100	16	19	0	2	1	1

*Probably primary thrombosis of pulmonary arteries.
†Filter wrongly placed in iliac vein.

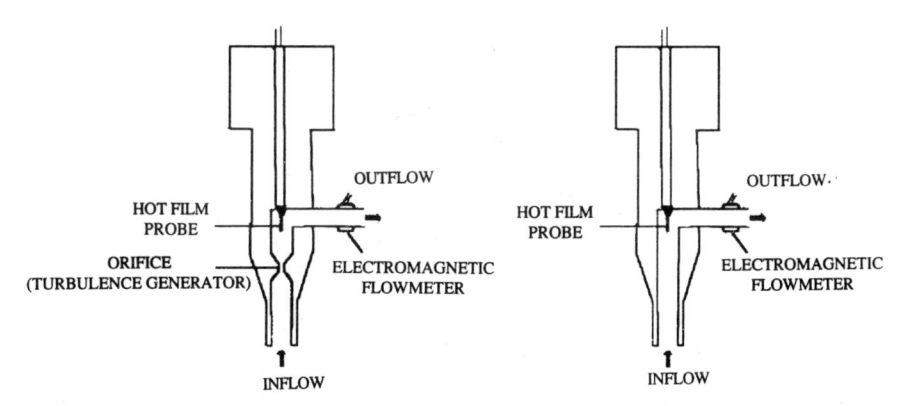

Fig 3.4 Diagram of the two shunts implemented by Stein and Sabbah (11). Left: turbulent shunt. Right: laminar shunt. (Copyright (1972) American Heart Foundation.)

(Fig. 3.4) between femoral artery and the contralateral femoral vein in each animal.

One shunt allowed straight laminar flow, while the other produced turbulence of a magnitude already quantified in previous *in vitro* investigations. Twenty-one separate studies were performed on each dog and the results gained lead Stein and Sabbah to conclude:

> Turbulence appears to contribute to the formation of thrombi.
>
> More thrombi were recovered from the turbulent system than the nonturbulent system, and the amount of thrombus formation (weight of thrombi) seemed to be related to the relative intensity of turbulence.

The MU filter was superseded (in clinical practice) by the Kimray–Greenfield (KG) filter (12)–(15) and this newer type of filter ironed out some of the problems found by the MU filter.

In 1973, a report produced by Greenfield *et al.* (12) looked into the insertion of a cone shaped wire filter (Fig. 3.5) into twenty-four dogs. The dogs were split into two groups. The first group had filter insertion only, for periods of 1–12 weeks. The second group were injected with 10 ml of thrombus immediately following insertion, and follow up was performed for 6–16 weeks. Experimental clots are formed by the addition of various enzymes or inhibitors to a blood sample, with regulation of size controlled by the area and volume of the mixing container.

Venocavagrams were made on the dogs at weekly intervals and no immediate or late complications occurred, all thrombi were trapped and the patency of the filter was maintained.

The report also discussed the major advantage of the KG filter over the Mobin-Uddin filter, being that once emboli are trapped in an MU filter, its gaps can become occluded and turbulence increased, thus causing an increase in the

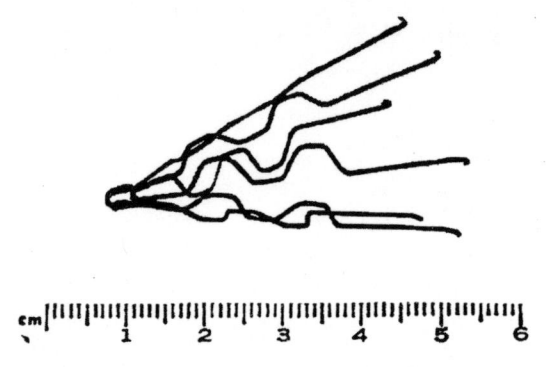

Fig 3.5 Side view of the intracaval device used by Greenfield *et al.* (12), shown in fully expanded position

possibility of thrombosis. However, the KG filter could be filled up to 80 percent of its depth with only a 64 percent of its cross-section impaired, thus reducing intraluminal disturbance (Fig. 3.6).

In 1985, Messmer and Greenfield (**16**) produced a long-term follow up of 235 patients inserted with the KG filter. The study group consisted of sixty-nine patients with a mean insertion time of approximately four years.

The results gained were split into four areas: filter position, span, migration, and angle. The results are described diagrammatically in Fig. 3.7.

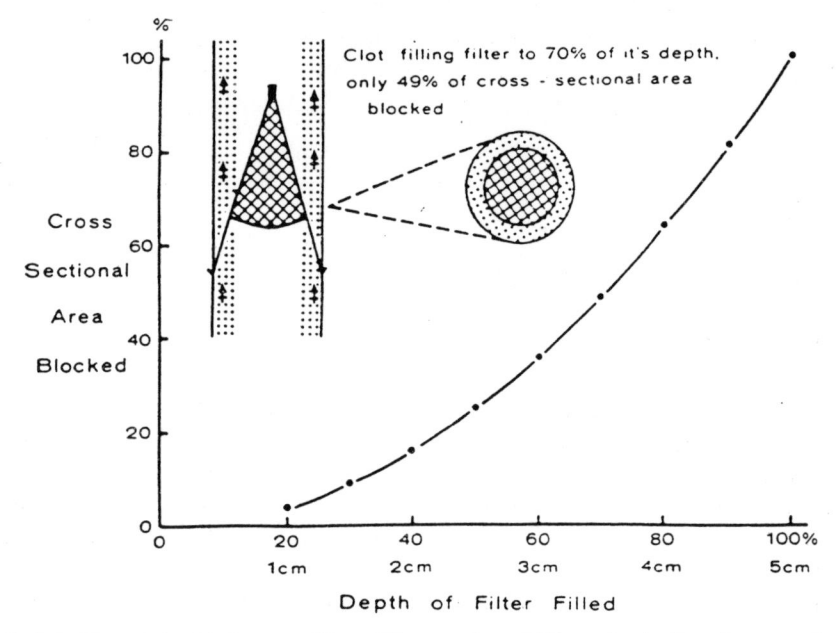

Fig 3.6 The relationship between filling of the cone-shaped filter and reduction of cross-section as discovered by Greenfield *et al.* (12)

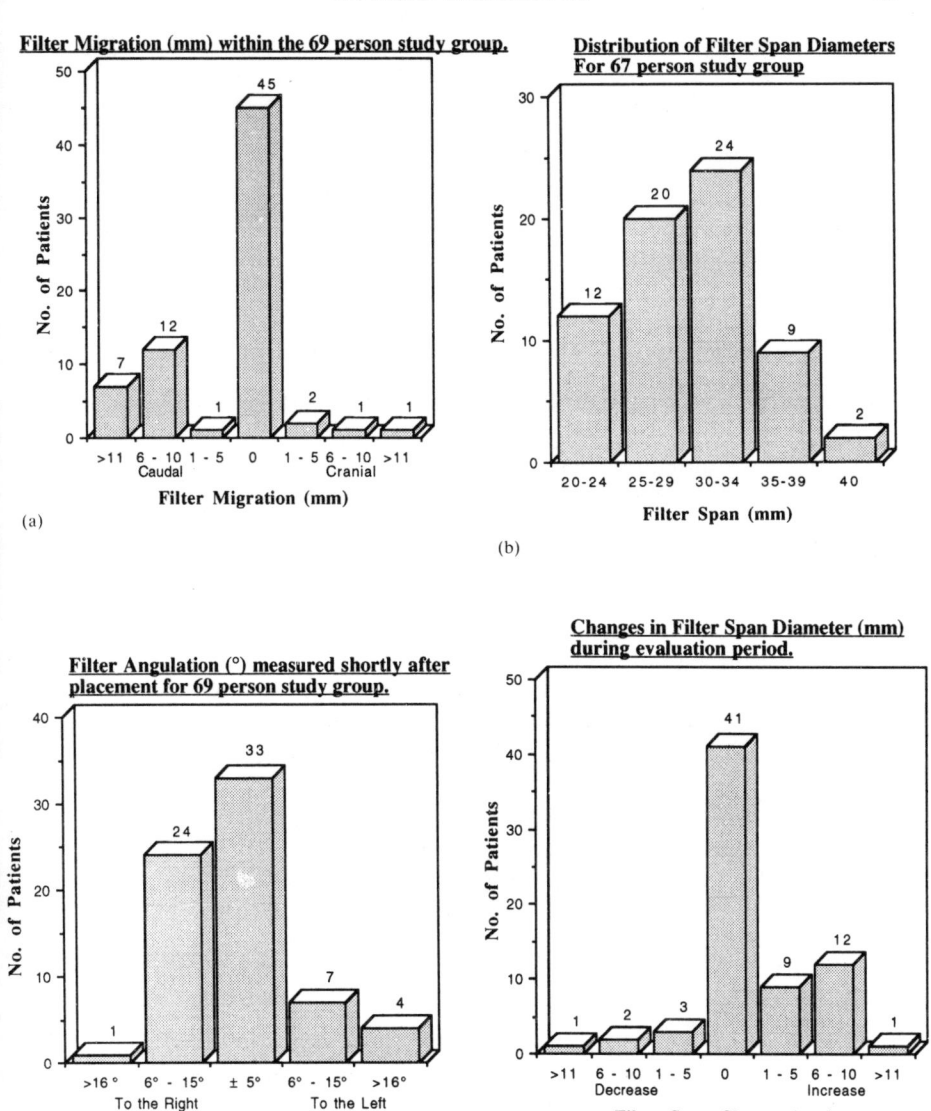

Fig 3.7 Graphical results gained by Messmer and Greenfield (16)

Although the results from this study were very encouraging, the KG filter still has some problems and disadvantages (**17**):

(a) It uses a cut down procedure of a major vein requiring dissection of either neck or groin, of an already ill patient, so that the relatively large delivery capsule can be placed through a venotomy.

(b) The orientation of the filter is difficult to control on delivery within the inferior vena cava (IVC) and, unless the apex of the filter is centralized, the efficiency of the filter may be reduced.

An ideal filter on the other hand would have to be able to do the following:

(a) Be deliverable into the IVC using a standard angiographic catheter.
(b) Be able to be delivered quickly, accurately, and have no problem of wrongful orientation.
(c) Lock itself into position and not migrate heartward.
(d) Capture all dangerous emboli.
(e) Be non-thrombogenic and have good biocompatibility.

Even with its disadvantages the KG filter is still used but has become more and more under pressure by the plethora of new filters.

The Simon nitinol filter (SNF) is a new filter that hopes to improve further upon the vena cava filter success. The SNF is a Ni–Ti shape memory filter, and has been developed over many years.

3.2 The development of the SMA inferior vena caval filter

Simon *et al.* (**18**) undertook a preliminary investigation in the mid 1970s. The group used two test systems, one *in vitro* and the other *in vivo*, to look at the clot capturing capability of the basic Ni–Ti alloy filter (Fig. 3.8).

The initial filter design constructed by Simon *et al.* (**18**) consisted of a locking system and filter mesh.

The locking system incorporated two elements: sharp tipped leading and trailing wires which penetrated the endothelium to a depth of 1–2 mm, and were prevented from further penetration by small metal studs (Fig. 3.8).

Initially the mesh designs consisted of one of three types (Fig. 3.9):

(a) clover leaf (three, four, five leaves)
(b) spiral
(c) overlapping ring pattern.

Later investigations, however, used a criss-cross grid design (Fig. 3.9) that incorporated a more uniform opening size.

The filter wire was straightened in ice cold water and placed within its delivery catheter with the rear stud of the locking system placed within a special notch (Fig. 3.10). This enabled the rear end to be held until all the filter had been guided out of the catheter. The placement was then performed using the perfusion of cool saline to avoid the filter forming within its delivery system.

The *in vitro* test system used a latex tube to simulate a canine vena cava and had the flow rate and saline temperature regulated to correspond to the nominal canine conditions. Emboli of various sizes were then placed in the system and the filter's capture success assessed. Fifty filters were tested in the *in*

Fig 3.8 The basic Ni–Ti filter as used by Simon *et al.* (18). NB: Small studs prevent deep penetration. (Courtesy of RSnA.)

vitro system. Initial designs permitted significant emboli to pass through the system.

However, by adaption of the filter mesh to the aforementioned criss-cross design, this problem was alleviated and the size of emboli obtaining passage was reduced.

The *in vivo* system involved the introduction, via a cut down procedure of the left femoral vein, of the filter into the vena cava, just below the level of the renal

Fig 3.9 Experimental filter mesh designs used by Simon *et al.* (18), cloverleaf, grid, overlapping circles. (Courtesy of RSnA.)

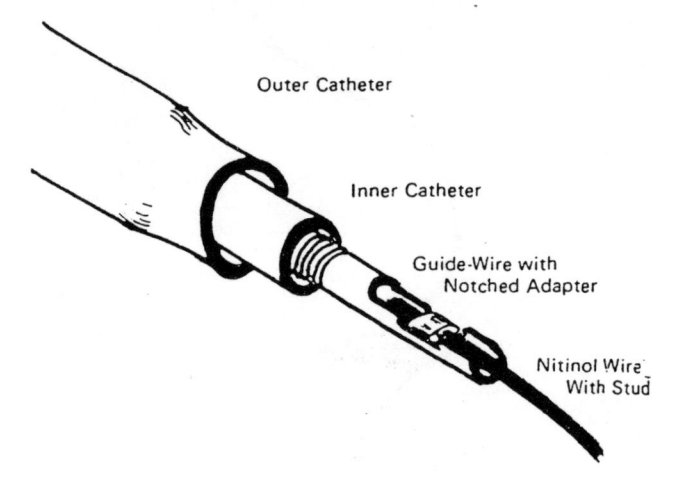

Fig 3.10 Diagram of delivery system for filter used by Simon et al. (18). (Courtesy of RSnA.)

veins in sixteen dogs. The emboli were then introduced via the right femoral vein using venotomy. The results highlighted problems with the basic design, hooks not forming properly in small vena cava and penetration of the walls of the vena cava. But the results also showed that the Ni–Ti filter became 'biologically' incorporated into the vena cava system, allowing neointimal growth over its area.

Simon et al. concluded that the 'initial experiment had proved very encouraging' and that investigation must be carried out to 'develop an optimum filter and delivery system suitable for human applications'.

In 1982 Palestrant, Prince, and Simon (**19**) reported on *in vitro* test comparison between a Ni–Ti alloy filter, the Mobin-Uddin filter and the Kimray–Greenfield filter.

Palestrant's *et al.* Ni–Ti filter design (**20**) had advanced from the simplistic design used by Simon *et al.* (**18**). The filter now consisted of seven mesh loops (outer diameter 25 mm) which were capable of trapping all emboli of 5 mm or above and a large proportion of smaller clots. Although Palestrant did not discuss the significance of size, it can be easily seen that the smaller the emboli allowed through, the less risk there is to the patient. The device had six anchoring legs that penetrated the endothelium layer by a depth of 1 mm and ensured optimal positioning of the filter within the lumen. This new design of filter became known as the Simon nitinol (SN) filter.

Figure 3.11 shows a comparison between the three filters and their delivery systems.

Palestrant's *in vitro* system (Fig. 3.12) comprised of a piece of transparent dialysis tubing housed in a Plexigas chamber, a pump that provided a regular flow from a reservoir of saline solution, and a thermostatically controlled heater to maintain the saline at body temperature.

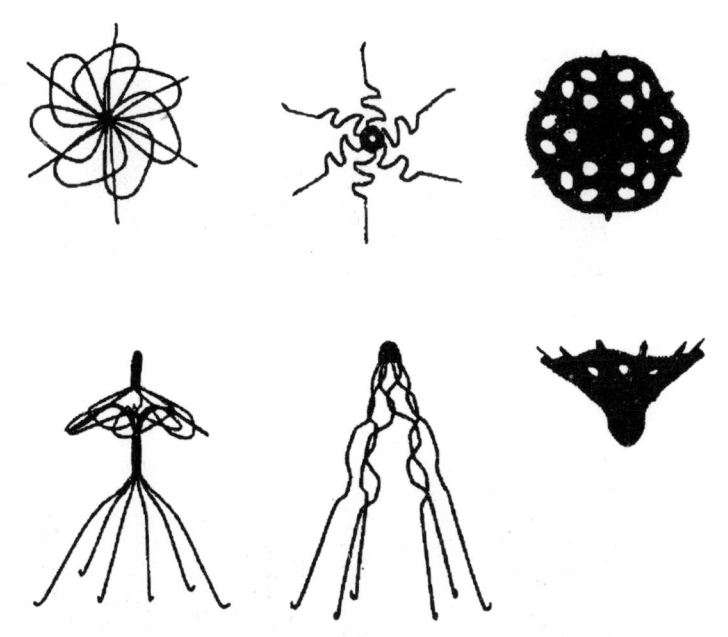

Fig 3.11 The three types of filters and their delivery systems compared by Palestrant *et al.* (19). (Courtesy of RSnA.)

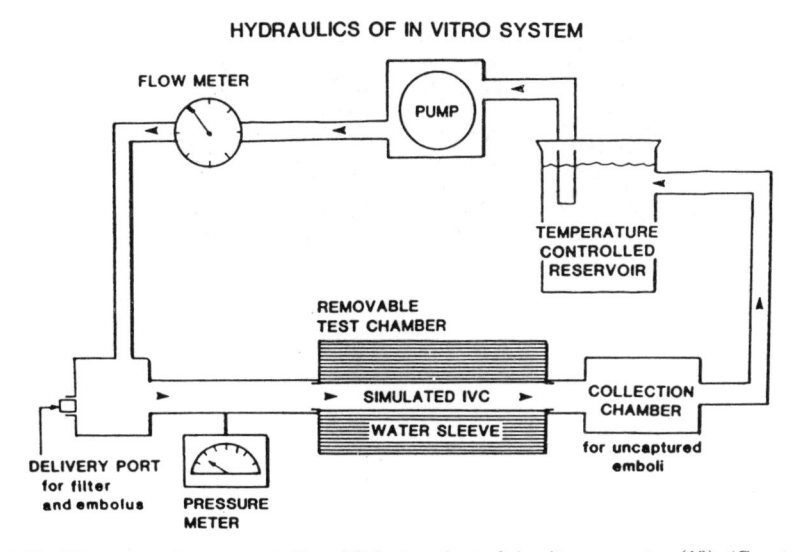

Fig 3.12 Diagrammatic representation of Palestrant's *et al. in vitro* apparatus (19). (Courtesy of RSnA.)

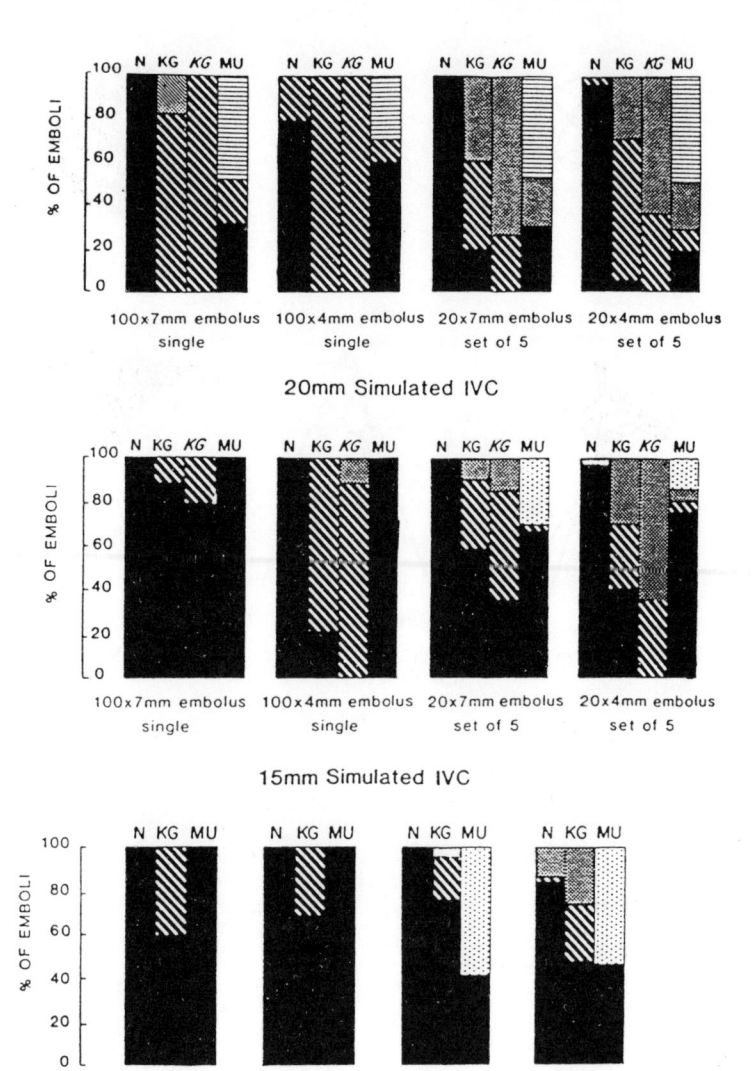

Fig 3.13 Effectiveness of embolus capture of the three types of filter used by Palestrant *et al.* (19). (Courtesy of RSnA.)

Table 3.2 Comparison of the three filters used by Palestrant et al. (19)

Filter	Embolus capture (% of large emboli)	Embolus capture (% of sheet emboli)	% of filter displacement§	Average pressure change (cm water)‡
Ni–Ti	100	97	0	<5
KG central	97	77	0	<5
KG tilted	97	54	0	<5
MU	87	49	0* 44†	19

*Inferior vena cava diameter of 15–20 mm
§The percentage of filters that moved after placement in either the anterior or posterior direction.
   For MU filter two differing percentages are used depending on caval diameter
†Inferior vena cava diameter of 28 mm
‡The increase of water pressure due to the trapping of emboli within the filter

Experiments were performed with three sizes of artificial vena cava, 15, 20, and 28 mm, representing the varying sizes found in humans. A side arm allowed delivery of the various filters into the test region. The results are shown for visible ease in Table 3.2 and Fig. 3.13.

The experimental results showed a significant difference in the performance of the three filters relative to embolus capture, tendency to migrate, and venous pressure changes.

Both the clinically used filters showed different disadvantages compared to the SN filter:

The Mobin-Uddin filter

(a) Exhibited marked rises in hydrostatic pressure and hence diminished flow once an embolus was trapped by it.
(b) In large diameter artificial IVCs the MU filter showed a tendency to migrate towards the heart.
(c) Its percentage embolus capture was less than that of either the KG or SN filter.

The Kimray–Greenfield filter

(a) This showed high captive rates as long as the apex of the filter was centralized within the lumen; once offset (or tilted) its capture rates dropped considerably and the possibility of passage by significant emboli increased.

Palestrant concluded that the SN filter was encouraging and showed potential for applications within humans.

In 1988 Katsamouris et al. (**21**) reported on an *in vitro* comparison of clot trapping and flow dynamics of a range of filters:

(a) Mobin-Uddin filter (MU) (**22**) – initially used and now clinically obsolete.
(b) Kimray–Greenfield (**17**) – clinical standard at time of report.

(c) Amplatz spider (A) **(23)** – new design.
(d) The Gunther basket (G) **(24)** – new design.
(e) The Simon nitinol (SN) filter – new design.
(f) The bird's nest (BN) filter **(25)** – new design.

Katsamouris' *et al.*'s *in vitro* system (Fig. 3.14) comprised of a reservoir and a circulating pump. The water in the system was at room temperature except for the SN filter where the water temperature was raised to 32.5°C to allow complete shape recovery to occur. A constant (pulsatile) flow of 2.0 l/min circulated round the set up.

A 27 cm long tygon tube, whose internal diameter was 23 mm, simulated the IVC and clots were introduced via a side arm in the system. The hydrostatic pressure was measured both sides of the filter. Eight different sizes of embolus were used: $2 \times 10, 3 \times 10, 3 \times 20, 3 \times 50, 4 \times 6, 7 \times 20, 6 \times 50$ and 6×90 mm and for each filter ten clots were introduced at one time. The filters were in two positions:

(a) Central or optimal position where all filters were ideally placed.
(b) Eccentric position. This meant tilting filters or placing them at non-optimal positions. Note that for the BN filter both elongation and compaction of the filter were carried out.

The results are shown diagrammatically in Figs 3.15 and 3.16 and showed the following:

(a) That the G, BN and SN designs would give more protection against pulmonary embolism because they do not tilt **(23)**.
(b) The best pressure gradients were obtained from the BN and SN filter.
(c) That insertion of BN, SN, A and G filters only created minor disturbances in laminar flow.

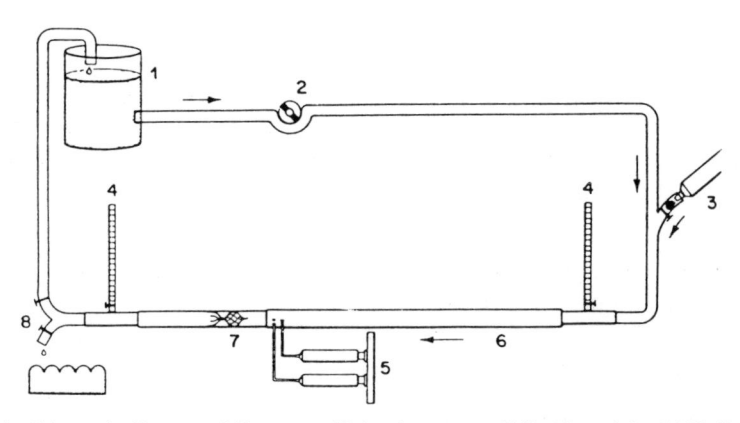

Fig 3.14 Schematic diagram of Katsamouris' *in vitro* system (21). (Copyright (1988) Springer-Verlag.)

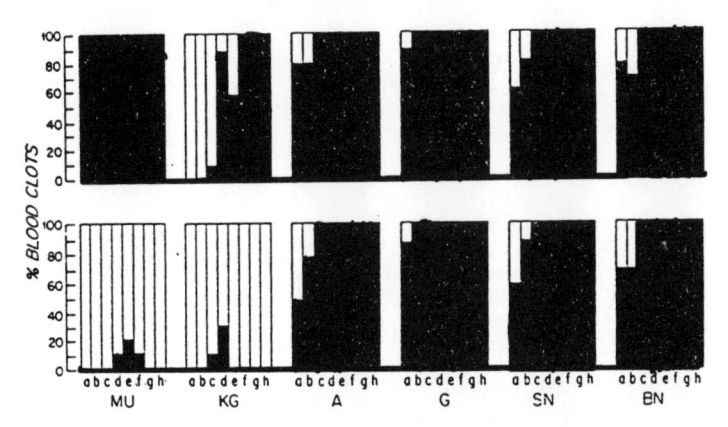

Fig 3.15 Percentage of clots trapped (shaded) and clots passed through (unshaded) in Katsamouris' *in vitro* system (21). (Copyright (1988) Springer-Verlag.)

It was concluded by Katsamouris *et al.* that:

SN, G and BN filters are superior and have most promising designs. The G and BN filters have been implanted in patients and have problems with migration (26)(27).

Around the time when the Katsamouris *et al.* report was being put together, the SN filter started in multi-institutional clinical trials and in 1989 Simon *et al.* (28) reported on their preliminary findings. The SN filter was tested in

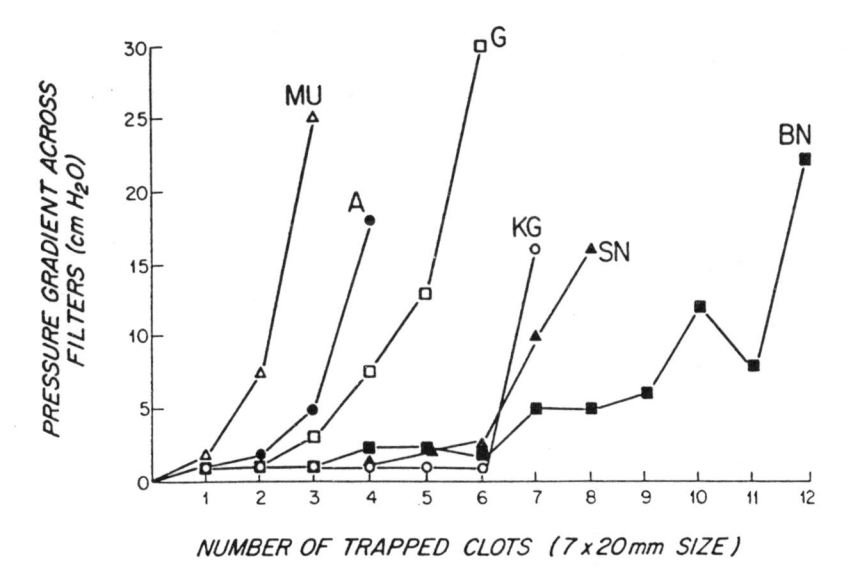

Fig 3.16 Graph showing the pressure gradient increase in relation to number of trapped clots (7 × 20 mm) in Katsamouris' *in vitro* system (21). (Copyright (1988) Springer-Verlag.)

Table 3.3 The summary of preliminary results gained from clinical trials of 103 patients in 17 centres (28)

Results	No. of patients
Migrations	
Cranial	0
Posterial	0
Recurrent pulmonary embolism	
Symptomatic	0 (1)
Asymptomatic	2
IVC occlusion	
Symptomatic	7 (2)
Asymptomatic	3
Deaths related to filter	0

Note: Numbers in parentheses represent additional patients with clinically suspected finding only.

seventeen centres across the United States and, at the time of the report, 103 filters had been placed within patients. Of the 103, none of the filters showed migration and there were no deaths related to the filter. The preliminary results are as shown in Table 3.3.

A more in depth study was carried out on a group consisting of forty-four people – twenty-six male and eighteen female, ranging in age from 19–96 years with the median being sixty-two years. In this study forty-one filters were implanted via the femoral vein (twenty-nine in the right and twelve in the left) and three were delivered through the internal jugular (two right and one left). Computer tomography was performed on three patients, and twenty-five magnetic resonance imaging (MRI) examinations (29) were performed on twenty-one patients. All further examinations are summarized in Table 3.4.

The results gained from the various tests are shown in tabular form in Table 3.5.

Simon *et al.* concluded that:

Initial experience with the SN filter is encouraging the procedure, for percutaneous filter insertion is simple, quick and relatively free of significant

Table 3.4 Summary of further examinations, shown at specified periods of time for the SN study group (28)

Test	No. of patients (initial test)	Soon after placement	Three months	Six months
Abdominal radiograph	44	–	7	4
Ultrasound study of entry site	–	18	5	2
Ultrasound study of filter region	–	9	5	2

Table 3.5 The results gained from SN study group (28)

Results	No. of patients
Perioperative complications	
Entry site of thrombus	5
Haemorrhage	0
Failure to introduce	0
Major displacement	0
Leg penetration	0
Migration	0
Tilting of dome*	24
Spindle formation†	1
Leg crossing‡	4 (2)
IVC patency	
Patient with thrombus	5 {11}
Occlusion (symptomatic)	3 (2) {7–11}
Occlusion (asymptomatic)	3 {7}
Recurrent pulmonary embolism	
Symptomatic	0 (1)
Asymptomatic	1
Deaths	
Related to filter	0
Unrelated to filter	7

Note: Numbers in parentheses represent patients with clinically suspected findings only. Numbers in curly brackets are percentages.
*Dome tilting of up to 30 degrees occurs as a normal adaptation to a small IVC lumen. It does not affect filtering efficiency, since the dome hole sizes are unaffected.
†Spindle formation is an intermediate stage of dome recovery that may occur in very small vena cava.
‡Leg crossing occurred in the first four patients due to a pusher wire problem that has been corrected.

complications of the veni puncture site due to the small size of the delivery catheter.

In the larger series of 103 patients, symptomatic occlusion occurred in 7–9 percent, a finding comparable to those in published series of other filters.

In the detailed BIH/MGH series, are demonstrated filling defects or

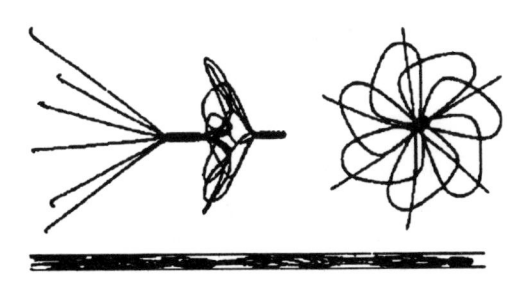

Fig 3.17 Advanced SNF used by Simon *et al.* (1989) (28). (Courtesy of RSnA.)

occlusion in 25–29 percent (eleven plus a possible two out of forty-four patients), comparable to 39 percent in a contrast material – enhanced CT study of another filter.

3.3 Recent research into the SMA inferior vena cava filter

All the previous reports highlighted so far have discussed the percutaneous introduction of the SN filter via either femoral or internal jugular veins. However, the use of the femoral vein may be constrained by the effects of deep venous thrombosis and the internal jugular route may cause complications. In 1990, Ducksoo et al. (**30**) produced a report appraising the use of external jugular entry as a possible alleviator of such problems. The group successfully placed five filters into the inferior cava via the right ($n = 4$) and left ($n = 1$) external jugulars. However, only one of the five devices used was an SN filter, which accessed the IVC via the left external jugular.

The external jugular runs at an angle down the neck, where it eventually connects to the subclavian vein (Fig. 3.18). From here entry into the superior vena cava, and hence to the IVC, is achieved.

Ducksoo et al. concluded that the external jugular vein was:

> a safe alternative access route for placement of the new inferior vena caval filters when access via the femoral vein or internal jugular is impractical.

The group also commented on the SN filter's ability to be placed via the more difficult left external jugular approach, and stated that this route would be more complicated by other filters with:

> somewhat larger and stiffer introducers.

In a further report in the same year, Rose and Miller (**31**) discussed improvements in the control of SN filter delivery.

In several papers (**31**)(**32**)(**20**)(**33**) the rapid restructuring of the filter has been reported to cause what is sometimes termed as a caudal 'jump' (Fig. 3.19). This has meant that the filter is positioned several centimetres lower than desired. On most occasions this 'jump' does not cause any serious complications; however, avoidance of the problem would be beneficial.

Rose and Miller used an iced saline perfusion technique, where the amount of saline used was around 200 ml in total. This was an increase on the recommended refrigerated saline technique as described by Nitinol Medical Technologies Incorporated, the manufacturers of the SN filter.

The group concluded that:

> Cooling the saline to freezing point, increasing the perfusion rate, and diminishing the transit time allowed for a controlled two step filter expansion, and assurance of optimal filter position within the inferior vena cava.

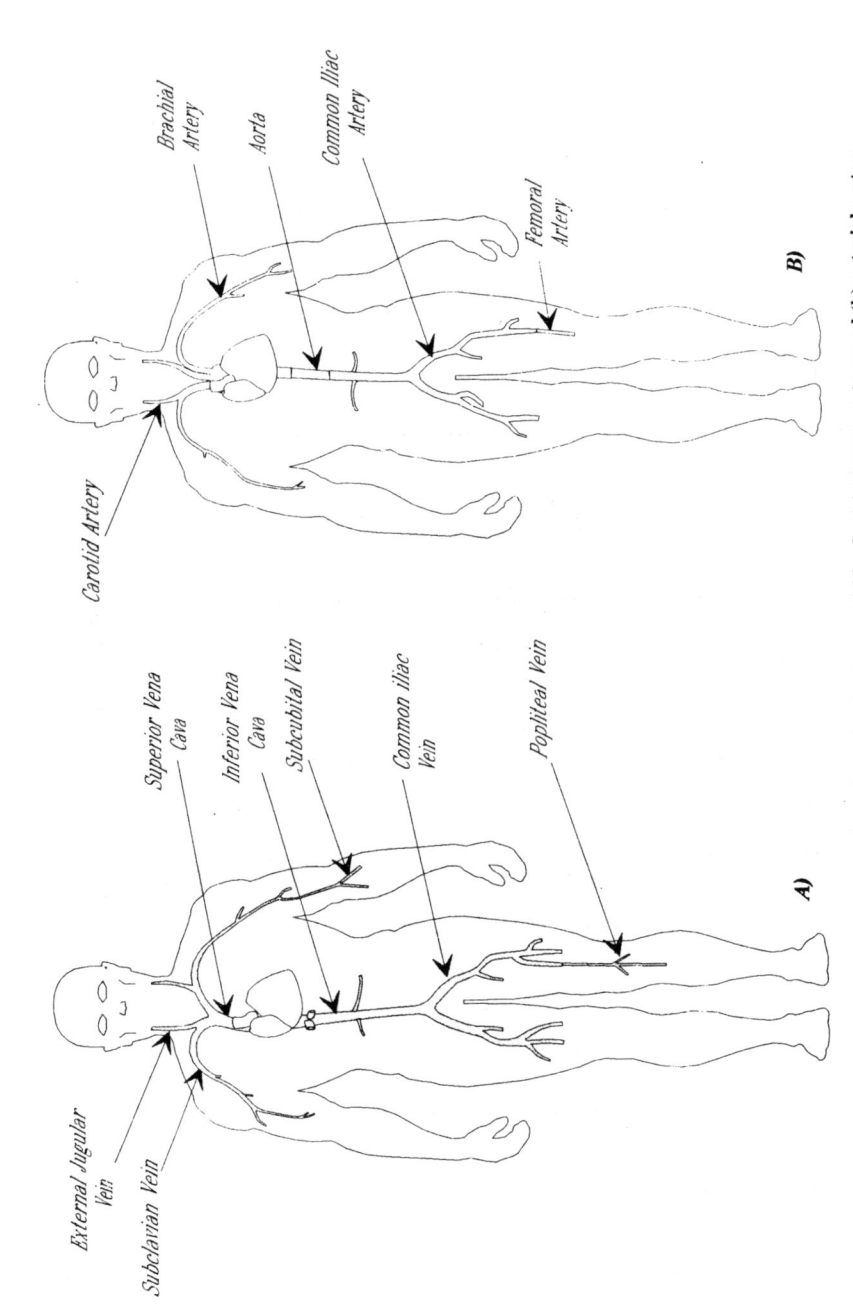

Fig 3.18 Highly diagrammatic representation of vessels of interest within the (a) venous system and (b) arterial system

46   The Application of Shape Memory Alloys in Medicine

1.

2.

Inner Vessel Wall (Intima).

Pusher Wire.

3. As the filter mesh extends and reaches the boundaries imposed by the intima a strain builds up within the mesh.

At this point the filter legs have not been released from the catheter and thus the filter is not locked in the desired position,

The strain is transformed into a posterial movement and hence a slight caudal 'jump' away from the placement site occurs.

4. Once the filter legs have extended beyond the placement catheter, they reform their memorized shape and lock the filter in position within the IVC.

Fig 3.19   A diagrammatic representation of a caudal 'jump'

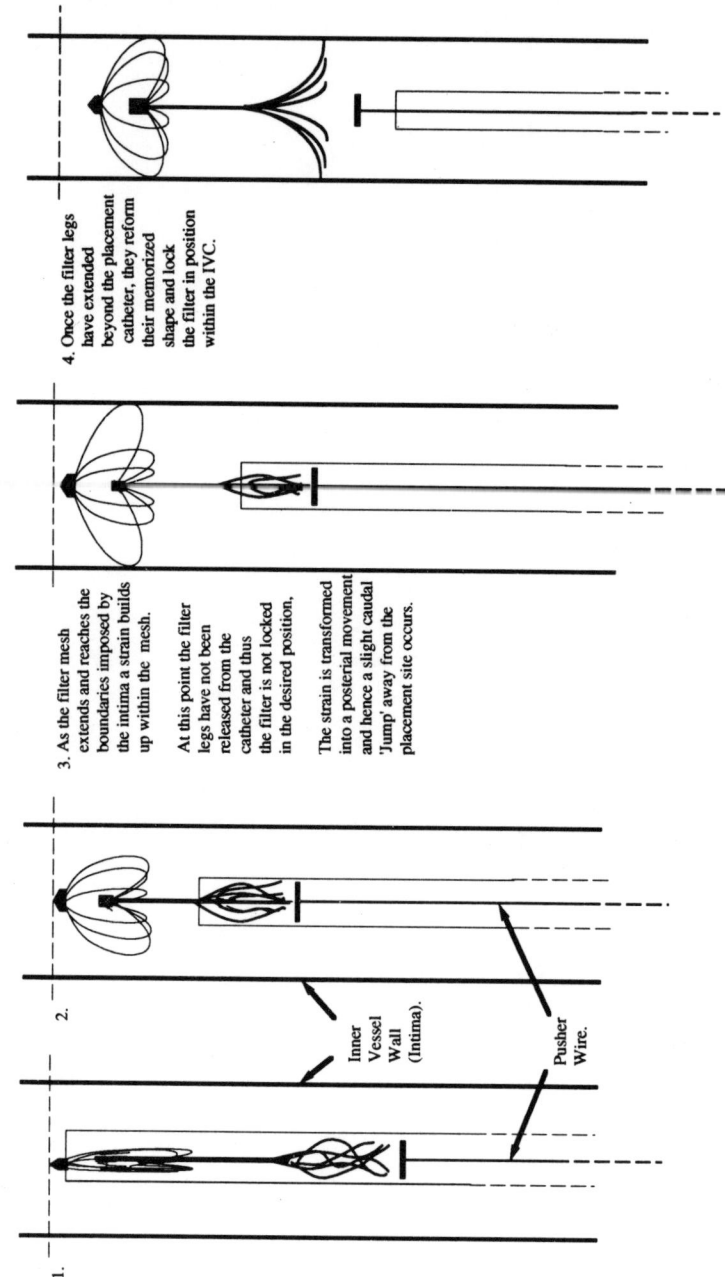

Table 3.6 Physical characteristics and clinical availability of seven caval filters (34)

Filter	Material	Length (mm)	Diameter (mm)	Outer diameter of delivery mechanism (F)		Maximum caval diameter (mm)
				Carrier	Sheath	
Steel Greenfield*† (SG)	316L stainless steel	41	30	24	29.5	30
Bird's nest*‡ (BN)	304 stainless steel	70§	54	11	14	40
LGM*†	Eligiloy	46	30	10	12	28
Titanium Greenfield†¶# (TG)	Beta-3 titanium	47	38	12	14.3	30
Simon nitinol¶ (SN)	Nickel–Titanium	38	28	7	9	28
Gunther**†† (G)	304 stainless steel	75	22–30	10	12	28
Amplatz†††‡‡§§ (A)	MP32-N stainless steel	25	36	12	14.3	33

*Available in most major markets. †Filtration sensitive to tilting.
‡Second-generation filter currently marketed. Struts enlarged to diameter of 0.45 mm. Nest composed of 4 × 25 cm long, 0.18 mm wire.
¶Under trial.   §Ideal size.
#New hook design to prevent caval penetration; trials currently being performed.
**Available in most major markets except in the US and Canada.
††Retrievable.   ‡‡Trials complete.
§§May be manufactured in larger size for use in vena cava of larger diameter.

Dorfman produced a review (32) of all the IVC filters (Table 3.6, Figs 3.20 and 3.21) available at the time of publication (1990). The report discussed the morbidity and mortality of non-filter treatments and gave a brief summarization of the published effects of filter placement for all the devices highlighted.

Although a large amount of data were available, Dorfman concluded that:

> The most appropriate method to evaluate all the currently available devices would be a multicentre, randomized, prospective trial ... until then, it is mandatory that each of us review the available literature critically because the data from various studies are not directly comparable and, unfortunately, may often reflect the bias of individual investigators.

In another review article released in 1991, Grassi discussed the five Food and Drug Administration (FDA) approved filters (20) available for use in the United States (Fig. 3.22). Within his discussion, Grassi commented on the ease of application, the efficacy of each filter and the morbidity associated with placement. Grassi based his review on the published results submitted to the FDA and also called for randomized clinical testing to ensure fair comparisons to take place.

The report highlighted the deficiency in intravascular filters when attempting to prevent pulmonary emboli in oversized inferior vena cava (diameter >28 mm). With exception of the bird's nest filter, all show migration or capture difficulties and so are of limited application.

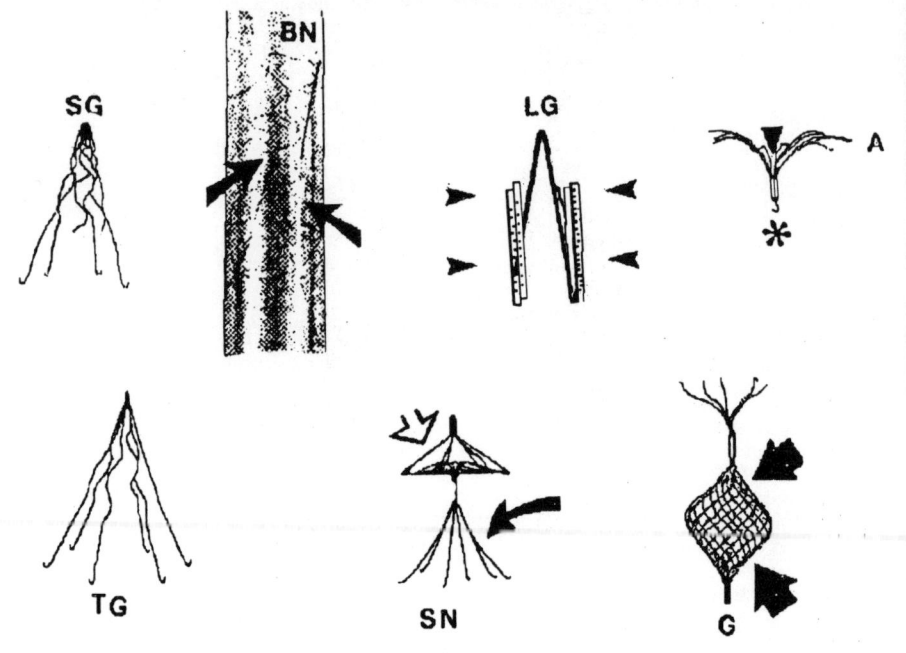

Fig 3.20 The seven inferior vena cava filters investigated by Dorfman (32), viewed from the side. (Courtesy of RSnA.)

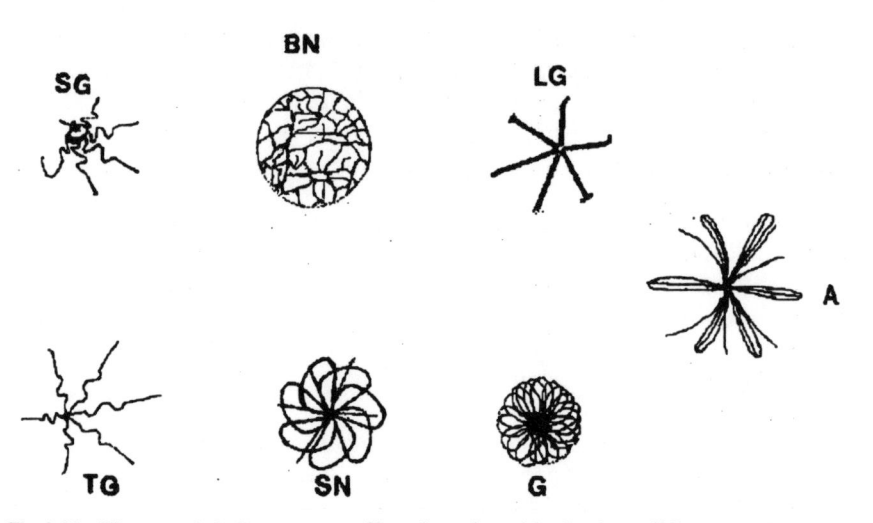

Fig 3.21 The seven inferior vena cava filters investigated by Dorfman (32), viewed from above. (Courtesy of RSnA.)

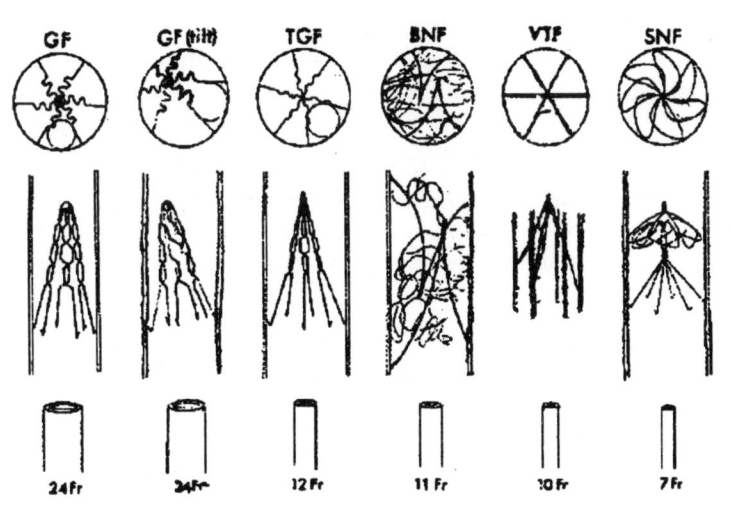

Fig 3.22 Diagram of implant filters used by Grassi *et al.* (20)

Table 3.7 describes the introducer dimensions, maximum recommended vena caval diameter, and the effect of tilting on the five filters.

However, within his paper Grassi did comment that the Simon nitinol filter was, in his own institution's experience:

> valuable when the smaller 9-French percutaneous puncture is advantageous, or where the longitudinal distance between the renal veins and iliac bifurcation is reduced.

Another report in 1991 by Kastan *et al.* (**33**) also covered the procedural difficulties encountered by clinical work with the SN filter. Kastan *et al.* also pointed out the caudal 'jump' but corrected this deficiency by a slight advancement of the pusher wire whilst unsheathing.

The group also noted a problem with the pusher wire itself; in 25 percent of cases tested by the group, the SN filter remained within the sheath, even after

Table 3.7 Comparison of five inferior vena caval (IVC) filters (35)

Filter type	Introducer size (OD) in French	Carrier size in French	Maximum IVC diameter recommended	Tilting significant?
Greenfield stainless steel	29	24	28	Yes
Greenfield titanium	14	12	28	Yes
Bird's nest	14	11	40	No
Vena-Tech	12	10	28	Yes
Simon nitinol	9	7	28 (2[a])	No

Note: OD = outer diameter.
[a]Manufacturer's recommendations before surgical procedures requiring general anaesthesia.

full extension of pusher and full withdrawal of sheath. This drawback was put down to manufacturing variations and, through communication of this problem with Nitinol Medical Technologies Incorporated, the pusher wire has been increased in length.

Even with these slight difficulties Kastan *et al.* concluded:

> Overall our experience with the SNF confirms that it is an easily placed percutaneous IVC filter. However, some technical difficulties were experienced in filter placement. In most cases, the problems resulting from these procedural difficulties were minor.

In a technical note published by Ducksoo *et al.* in 1991 (**34**), the antecubital vein approach to SN filter insertion was discussed.

In a similar fashion to that discussed by Ducksoo *et al.* (**30**), the antecubital vein shows possible advantages over the more conventional internal jugular or femoral routes. Unlike lower limbs, the upper limbs' venous blood is predominantly drained by a number of superficial veins.

The deep venous system of the arm can be reached with relative ease, and via the antecubital vein the placement of the SN IVC filter may be achieved. Ducksoo *et al.* concluded:

> When standard femoral or jugular approaches are not available, the Simon Nitinol filter offers several technical advantages.
>
> The antecubital route is an excellent alternative when conventional routes are compromised.

Ducksoo *et al.* also discussed, as they had with the external jugular vein approach, the decrease in possible complications arising from this new deployment route.

In 1992 Grassi *et al.* (**35**) published a report on the occlusion of vena caval lumen in patients with malignancy. Patients with malignancy often show a state of hypercoagulability (**36**). This increased risk of deep venous thrombosis, and hence of associated pulmonary embolism, means that placement becomes necessary if drug regimes fail or are constrained.

In this clinical test, twenty-four patients (eight men and sixteen women) with the average age of 60+ years were percutaneously implanted with the SN filter and findings taken over a thirteen month period. Grassi *et al.* found the SN filter to have a higher occlusion rate than was recorded in previous reports. But although this was the case, the group warned that no real significance could be taken from the findings until:

> a randomized, prospective trial with emphasis on follow up imaging of IVC is performed.

A further report in 1992 by Korbin *et al.* (**37**) discussed a direct *in vitro* comparison of four types of filters (Greenfield, Vena Tech-LGM (**38**), Simon nitinol, and Gunther) deployed within an artificial oversized lumen (phantom).

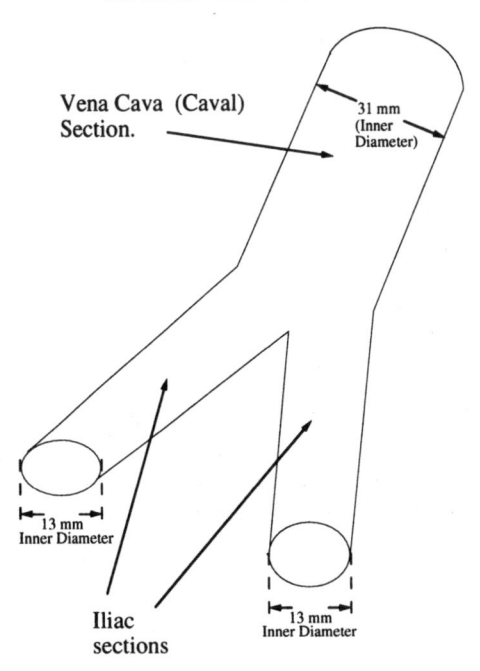

Fig 3.23 The inferior vena cava/common iliac oversized phantom used by Korbin *et al.* (37)

The phantom consisted of a glass Y-shaped tube with internal diameter of 31 mm at the IVC section and common iliac inner diameters of 13 mm (Fig. 3.23).

Laminar flow was encouraged through the system by the increased luminal length of each section.

A constant flow of 2 l/min was maintained through the system and pressure measurements were obtained. After filter placement blood clots were introduced to the system, Korbin *et al.* assessed the flow dynamics of the clots to produce stasis of flow, and a qualitative measurement of the turbulence produced for each filter. The results are shown in Table 3.8.

Table 3.8 The *in vitro* flow dynamics of the filters tested by Korbin *et al.* (37)

Filter position and type	Pressure gradient (cm H_2O)		No. of 7 × 25 mm clots needed to achieve fluid stasis	Dye stream turbulence with one 5 × 15 mm (quantative measurement) captured clot
	With no clots	With one 5 × 15 mm clot		
Caval segment				
BNF	0.8	0.84 (+5%)	8	Minimal
Iliac segment				
Greenfield	3.8	4.3 (+13%)	2	Marked
Vena Tech-LGM	4.1	4.5 (+10%)	2	Marked
SNF	3.9	4.4 (+14%)	2	Mild

Table 3.9 Diagnostic reasoning behind filter placement in the McCowan *et al.* study group (20)

Indication for filter placement	No. of patients
Deep venous thrombosis (DVT) above the knee as recurrent PE with anticoagulation constrained	12
DVT or recurrent PE when anticoagulation has failed or is complicated	4
Prophylactic placement: placement to prevent PE in high risk group though no other problems reported previously	4

Although concluding that the BNF filter was clearly preferable in an oversized IVC, the group pointed out that:

> The unpenetrable nature of the phantom walls did not allow normal seating of filter hooks and prongs. This may have led to altered filter device configuration, thus resulting in different flow and turbulence patterns of these devices than might be expected *in vivo*.

Again in 1992, McCowan *et al.* (39) published a paper looking at complications found with the SN filter.

The group used a small study of twenty patients; ten men and ten women, who had a mean age of sixty-five and had an average follow up time of fourteen months, with the reasons for filter placement shown in Table 3.9.

Four people were lost to follow up and for one person of the study group follow up involved only minimal evaluation. The results and complications encountered by McCowan *et al.* are tabulated in Table 3.10.

McCowan *et al.* commented on the high percentage of caval penetrations occurring in the study group. They noted the problem of caudal 'jump', as previously described. They also noted occurrence in two cases of the filter leg fracturing that, although were asymptomatic, had no definite cause. However, McCowan *et al.* did conclude that the Simon nitinol filter:

> appears to provide adequate protection from clinically significant PE. The small size of the introducer catheter can be an advantage in selected cases.

Table 3.10 Results gained by McCowan *et al.* (39)

Result	No. of occurrences	No. in study group
Pulmonary embolism	0	20
Caval thrombosis	2	20 (10)
Caval penetrate	5	20 (25)
Caval stenosis	1	20 (5)
Filter migration	1	20 (5)
DVT	1	20 (5)

Note: Numbers within parenthesis are percentages of the overall study population.

The SNF deserves continued close investigation with short and long term follow up to assess its biomechanical stability, effectiveness and safety.

Hawkins and Al-Kutoubi (**40**) reported on the initial SN filter experience in the UK. They used a group of seven patients with a range of ages between sixty-one and eighty-three. The indications and results obtained by the study are shown in Table 3.11.

Hawkins and Al-Kutoubi reported that there were:

No local complications at puncture site and no cases of PE occurred. There were no cases of filter misplacement.

However, the group did mention the problem of the caudal 'jump', but concluded that:

Our preliminary experience with the SNF has been encouraging. We have been impressed by the simplicity and rapidity of insertion and lack of symptomatic complications.

We now use the SNF preferentially and reserve the bird's nest filter for patients with an IVC of greater than 28 mm, for which the SNF is not suitable.

In February 1993, a group headed by Jon S. La Plante (**41**) reported on two cases of cranial migration of the SNF.

La Plante *et al.* had placed 117 SNFs over two years and the group noted that the two problem filters had been deployed as expected within lumen of diameters less than the maximum recommended by the manufacturers. However, La Plante *et al.* commented that:

Despite the occasional occurrence of migration, we believe that the Simon nitinol filter is a very safe filter.

Table 3.11 Patient disposing factors and results after placement by Hawkins and Al-Kutoubi (40)

Patient	Age (years)	Indication for SNF and later assessment
1	61	Cirrhosis, deranged, clotting, varices, DVT. Asymptomatic at 10 months
2	74	Elective cholecystectomy, previous DVT/PE. Asymptomatic IVC obstruction with minor leg swelling at 6 months on anticoagulant
3	83	Iliac DVT, GI bleed on anticoagulant. Asymptomatic, stocking to limit unilateral leg oedema after 6 months
4	63	Recurrent DVT and PE on anticoagulant. Asymptomatic at 8 months, on anticoagulant
5	75	Iliac DVT, pre-operative for benign ovarian cyst. Asymptomatic at 6 months, on anticoagulant
6	83	Iliac DVT, longstanding immobility. Asymptomatic at 1 month, on anticoagulant
7	62	Recurrent PE on anticoagulant. Died with ARDS at 1 month. No filter-related complications

Table 3.12 Filter design used by **Becker et al.** (42) (Copyright (1992) American Medical Association.)

Filter	Diameter, F		Material	Design
	Inner	Outer		
Greenfield, Medi-Tech, Watertown, Mass	24	29.5	Stainless steel	Pyramid
Bird's nest, Cook, Bloomington, Ind	11	14	Stainless steel	Random tangle
LGM, LG Medical, Chassemeuil, France	10	12	Stainless steel	Pyramid with lateral struts
Titanium Greenfield, Medi-Tech	12	14	Titanium	Pyramid with redesigned hooks and smaller caliber
Simon-nitinol, Nitinol Medical Technologies, Woburn, Mass	7	9	Nickel Titanium	Pyramid and dome
Gunther†	10	12	Stainless steel	Elongated basket with struts
Amplatz‡	12	14	Stainless steel	Inverted pyramid with apical hook

†Still in development; not manufactured for commercial use.
‡Theoretically retrievable.

The group continued that:

> We have found it to be simple to deploy and versatile because of its softness when cooled; it also has excellent filtration capability.

An overview of the IVC filters, their safety, effectiveness and indications was released in 1992 by Becker et al. (**42**). They considered seven filters (Fig. 3.24 and Table 3.12): Greenfield filter, titanium Greenfield filter, Simon nitinol filter, LGM or Vena Tech filter, Amplatz filter, bird's nest filter and the Gunther filter.

Becker et al. used an eight methological evaluation technique (Table 3.13) when considering twenty-four published papers (Tables 3.14 and 3.15).

The group's report commented on the wide variety of results with no

Table 3.13 The eight methods of assessment used by Becker et al. (42)

Standard	Description
1	Adequate diagnostic evaluation
2	Adequate description of patients studied
3	Adequate description of patient assembly process
4	Adequate analysis for severity of thromboembolic disease
5	Adequate description of treatment
6	Adequate documentation of adverse outcomes
7	Unbiased surveillance of patients
8	Complete patient follow up

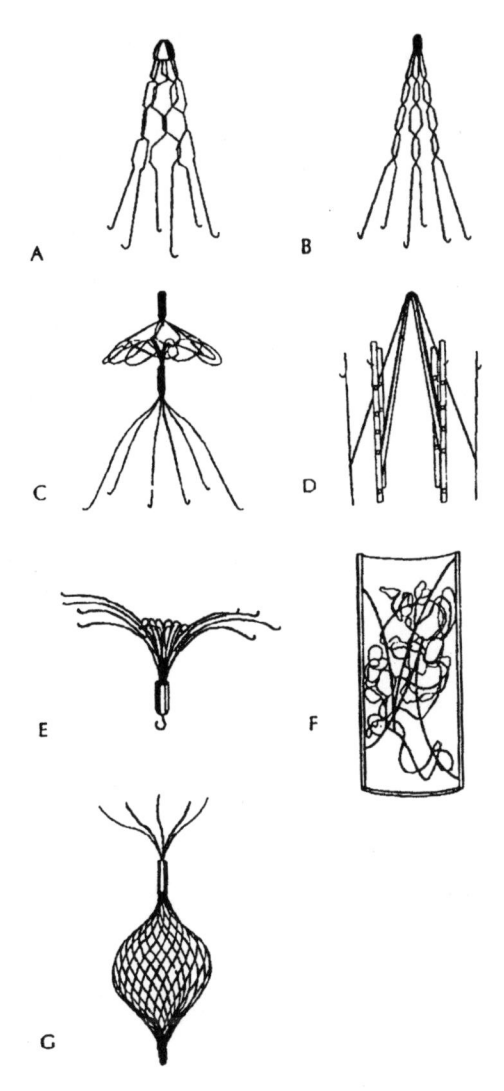

Fig 3.24 The seven filter designs investigated by Becker *et al.* (42). (Copyright (1992) American Medical Association.)

common standards of assessment. However, they did note that in the case of pulmonary embolism the mortality rate suggested that filters were:

> effective in preventing death from PE

and warned that:

> the indications for filter use are broadening before scientific evidence is available that filters are better than conventional prophylactic regimes.

Table 3.14 Summary of study methods as discussed by Becker *et al.*: Greenfield filters (42)

Source, year	Patient enrollment period	No. of patients	Type of study	Patients	Standards satisfied
Wingerd et al. (**43**), 1978	1970–1977	33	Consecutive series	Mixed	2, 3, 5
Cimochowski et al. (**44**), 1980	1974–1979	32	Consecutive series	Mixed	5, 6
Stewart et al. (**45**), 1982	1975–1981	12	Consecutive series	Mixed, suprarenal placement	2, 5, 6, 8
Scurr et al. (**46**), 1983	Before 1983	21	Consecutive series	Mixed	–
Swann et al. (**47**), 1986	1978–1982	9	Consecutive series	Neurosurgery	2
Carabasi et al. (**48**), 1987	1980–1986	200	Consecutive series	Neurosurgery, oncology	6
Rose et al. (**31**), 1990	1985–1987	109	Consecutive series, all percutaneous	Mixed	3
Whitney and Kerstein (**49**), 1987	1979–1984	10	Case series	Thrombocytopenia and cancer	5
Golueke et al. (**50**), 1988	1978–1985	88	Consecutive series	Mixed	3, 5, 6
Greenfield and Michna (**51**), 1988	1974–1986	469	Case series, several centres	Mixed	2, 5
Pain et al. (**52**), 1988	1986–1987	96	Consecutive series	Mixed	2, 3, 8
Todd et al. (**53**), 1988	1984–1985	21	Consecutive series	Mixed	2, 3, 5, 6, 8
Rohrer et al. (**54**), 1989	1978–1979	260	Consecutive series	Mixed	1, 2, 3, 6
Hye et al. (**55**), 1990	1984–1989	160	Consecutive series	Mixed	2, 5
Kolachalam and Julian (**56**), 1990	1982–1988	59	Case series	Mixed, 31 brain tumours, no anticoagulation	5, 8
Fink and Jones (**57**), 1991	1984–1988	42	Consecutive series	Mixed, no anticoagulation	2, 5, 8

However, the group concluded that:

> The available descriptive data suggest that the risk of filter placement for prevention of recurrent PE is justified in the face of anticoagulative contraindications or failure.

In a further comparative paper, Simon *et al.* (**64**) sought to evaluate five IVC filters – Greenfield (G), bird's nest (BN), Vena Tech (VT), Simon nitinol (SN) and titanium Greenfield (TG) – in physiologically simulated *in vitro* conditions. Simon *et al.* injected two sizes of clot into an *in vitro* system that was constructed to closely resemble conditions met inside a normal (21 mm) and oversized (28 mm) human IVC (Table 3.16).

Table 3.15 Summary of study methods as discussed by Becker et al.: newer filters (42)

Source, year	Country	Filter type	Patient enrollment period	No. of patients	Type of study	Patients	Standards satisfied
Castellani et al. (58), 1987	France	Modified Greenfield	1985–1986	35	Case series	Mixed	2, 5, 6
Febbe et al. (59), 1988	Germany	Gunther	1985–1988	59	Consecutive series	Mixed	2
Ricco et al. (38), 1988	France	LGM	1985–1986	100	Case series	Mixed	1, 5, 6
Roehm et al. (60), 1988	USA	Bird's nest	1982–1987	549	Multicentre case series	Mixed	–
Epstein et al. (61), 1989	USA	Amplatz	3y	52	Consecutive series	Mixed	2, 6, 8
Simon et al. (28), 1989	USA	Simon-nitinol	1988	44	Case series	Mixed	8
Greenfield et al. (62), 1990	USA	Titanium Greenfield	1988–1990	51	Case series	Mixed	5, 6, 8
Schneider et al. (63), 1990	Switzerland	Gunther	1987–1989	34	Consecutive series	Mixed	2, 5, 8

Table 3.16 Physical parameters simulated by Simon *et al.* (64)

Parameter	Human	Test system
Fluid media		
Composition	Venous blood, opaque	Dextran, NaBr Soln
Specific gravity	1.054 ± 0.06	1.06 ± 0.01
Viscosity	$4.5 \times H_2O$	$4.5 \times H_2O$
Temperature	37°C	37°C
Flow		
Pump	Heart	Cardiac bypass pump
Volume	1.5–3.0 l/min	Small cava, 2.0 l/min, large cava, 3.0 l/min
Velocity	10 cm/sec*	Small cava, 9.6 cm/sec, large cava, 8.2 cm/sec
Pressure		
Horizontal	5–10 cm H_2O	6 cm H_2O
Vertical	15–30 cm H_2O	20 cm H_2O
Pulsatility		
Rate	50–100 beat per min	Small cava, 40 beats/min, large cava, 80 beats/min
Amplitude	0–50 cm/sec	0–50 cm/sec
Turbulence	7.5–15 cm from iliac veins	12 cm from 'iliac veins'
Vena cava		
Composition	Thin, fibromuscular tissue	Thin, cellulose acetate
Size	15–35 mm diameter	Small cava, 21 mm diameter, large cava, 28 mm diameter
Shape	Round and oval	Round and oval
Orientation	Vertical and horizontal	Vertical and horizontal
Compliance	Soft, distensible	Soft, distensible
Filter position	Tip 15 cm above iliac veins	Tip 15 cm above 'iliac veins'
Clots		
Composition	Human blood clot	Human blood clot
Pliability	Soft, gelatinous, pliable	Soft, gelatinous, pliable
Size	Variable	Small, 2×30 mm; large 4×30 mm
Number	Variable	Sets of five delivered serially

*Approximate mean velocity.

A series of five clots of each size were assessed using four parameters; filter type, flow type (lamina, pulsatile), vena cava size, and orientation of the model (vertical or horizontal). Each series was repeated six times for a total of 240 emboli deliveries to each filter.

Simon's *et al.* results are shown in Table 3.17 and Fig. 3.25. They concluded that:

The SN filter proved the most efficient in the larger vena cava.

A comparison of clot capturing efficiency of five IVC filters available for clinical use in the United States used an *in vitro* model with eight sets of standardized test conditions. The overall random order for filter efficiency is 1: SN; 2: BN; 3: VT; 4: TG; and 5: standard G filters.

Fig 3.25 (on pp. 60–61) Results gained by Simon *et al.* (64) for: (a) horizontal IVC in 21 mm dia. cava; (b) horizontal IVC in 28 mm dia. cava; (c) vertical IVC in 21 mm dia. cava; (d) vertical IVC in 28 mm dia. cava. (Courtesy of RSnA.)

Table 3.17 Comparison by ranking (1 best, 5 worst) for filters used by Simon et al. (64)

Filter type	Horizontal*		Vertical*		Horizontal†		Vertical†		Total mean
	2 × 30 mm	4 × 30 mm	2 × 30 mm	4 × 30 mm	2 × 30 mm	4 × 30 mm	2 × 30 mm	4 × 30 mm	
G	5 (50)	5 (46)	5 (46)	4.5 (56)	5 (23)	3 (40)	4 (46)	4.5 (56)	4.50 (45)
BN	1 (86)	2 (80)	4 (50)	1 (70)	1.5 (90)	4.5 (33)	2 (83)	2 (83)	2.25 (72)
VT	3 (63)	3 (73)	2 (73)	4.5 (56)	3 (70)	2 (46)	3 (76)	3 (76)	2.94 (67)
SN	2 (76)	1 (86)	1 (90)	2.5 (60)	1.5 (90)	1 (90)	1.0 (90)	1 (90)	1.37 (84)
TG	4 (53)	4 (56)	3 (60)	2.5 (60)	4 (36)	4.5 (33)	4 (56)	4.5 (56)	3.94 (51)

Note: Numbers in parentheses are percentages of clots captured.
*IVC diameter, 21 mm.
†IVC diameter, 28 mm.

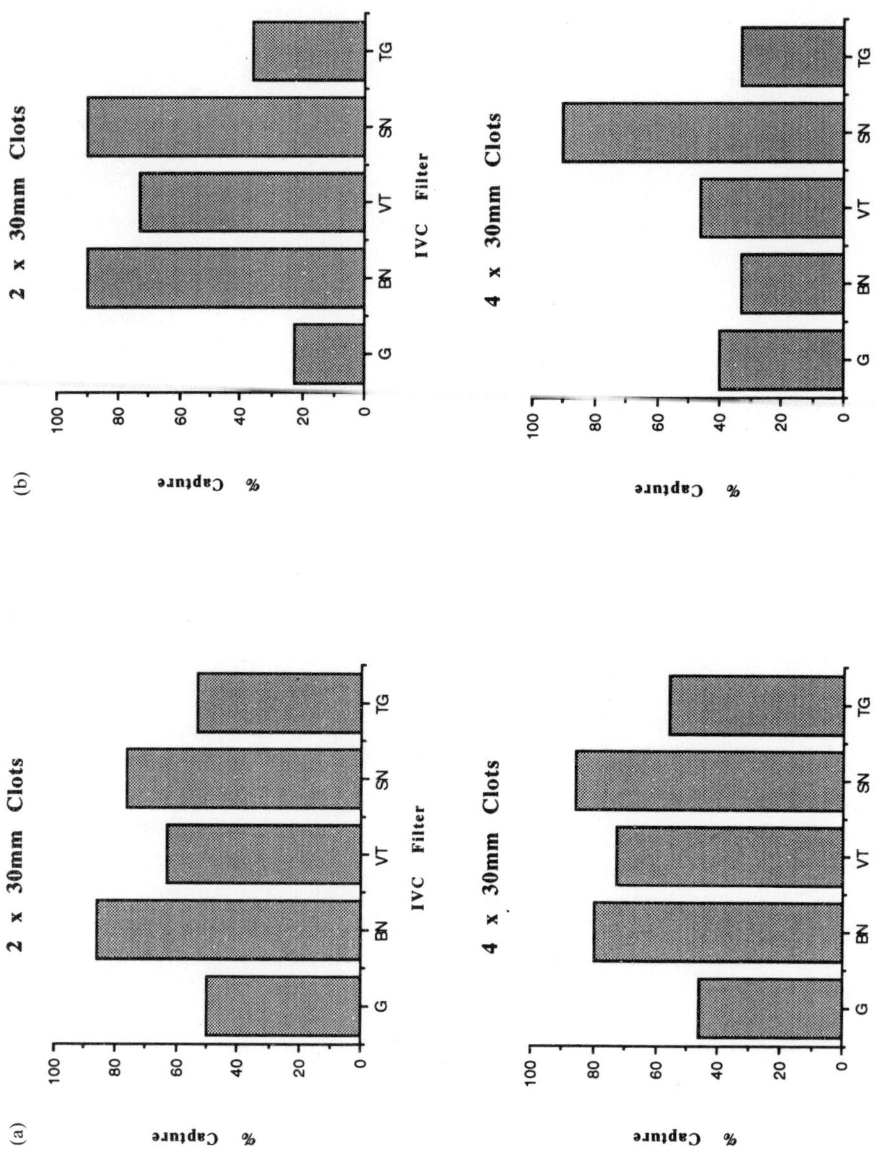
(a)
(b)
2 x 30mm Clots
4 x 30mm Clots
% Capture
IVC Filter
G
BN
VT
SN
TG
0
20
40
60
80
100

Fig 3.25

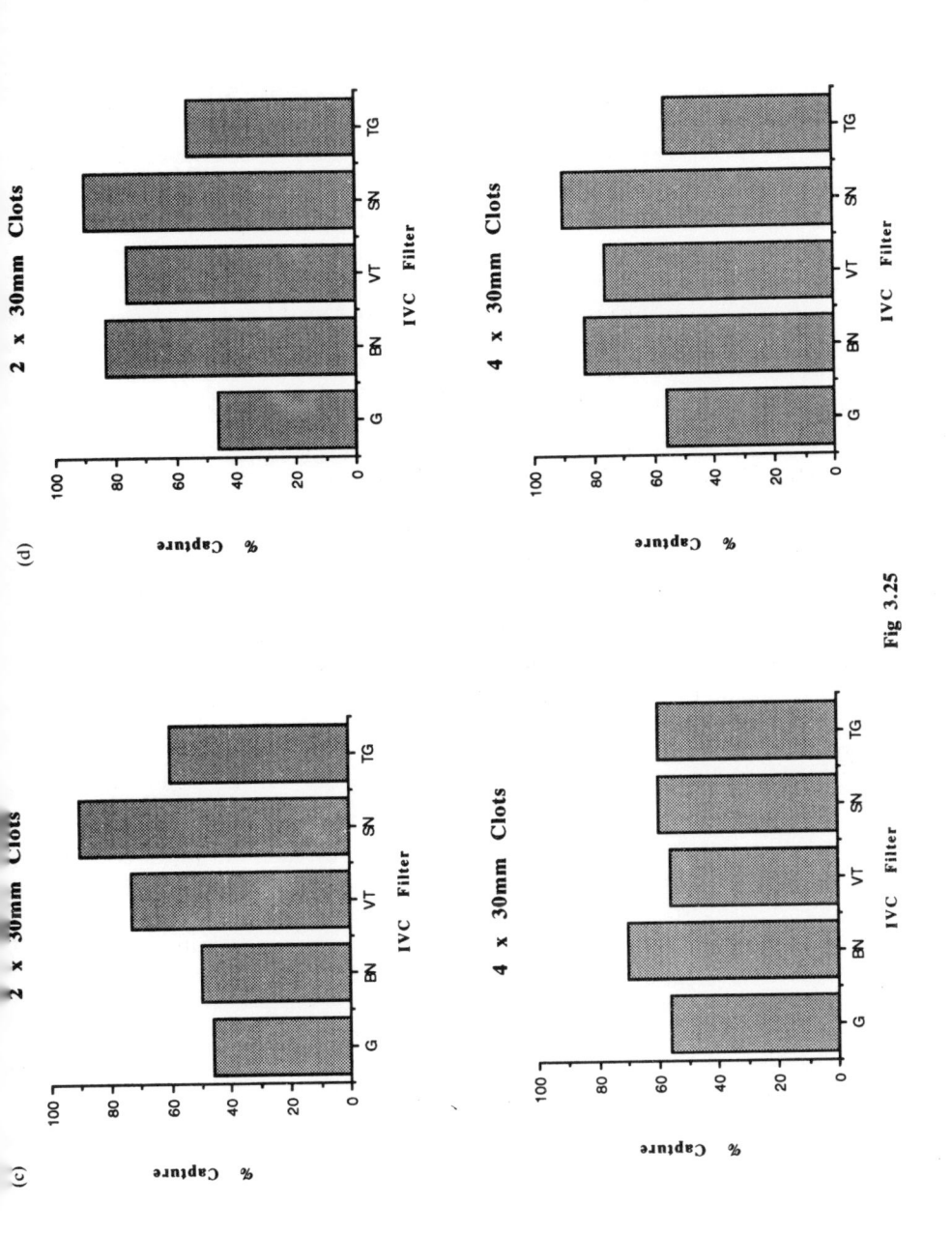

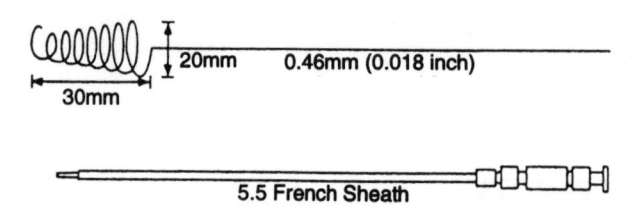

Fig 3.26 Nakagawa's *et al.* nitinol retrievable vena cava filter (65). (Courtesy American Heart Association.)

Whereas previous reports had concentrated on the long term patency of permanent SMA filters, Nakagawa *et al.* (**65**) investigated the experimental and initial clinical results of a retrievable nitinol IVC filter (Fig. 3.26). The group implanted filters into ten sheep and two patients, with retrieval occurring after a maximum of a week.

The group discovered that:

> The concept of filter retrievability is appealing since it may benefit patients on whom long-term IVC filters are not currently indicated.

Nakagawa *et al.* found no complications in either the placement or retrieval of all the nitinol filters and concluded that although further research was needed:

> The nitinol filter, when compared with permanent filters, has the additional advantages of easy surveillance and retrievability. It could expand the indications for prophylactic filter placement in several clinical conditions.

3.4 Conclusion

In recent years, percutaneous insertion of the Kimray–Greenfield (KG) filter (to create an interruption in the inferior vena cava) has been performed to prevent recurrent pulmonary embolism. However, this involves dilation of the veni puncture site to accommodate the 29F introduction sheath (approximately 9.6 mm in outside diameter (**66**)). Problems of pulmonary embolism have occurred if either the KG filter's apex is filled with a clot, or if its apex is not centralized within the lumen (**67**). Other difficulties such as migration, misplacement, puncture site thrombosis and fracture of the actual filter (**48**)(**68**) have also been observed.

No ideal filter has been designed; all show complications and no randomized multicentre test has been set up, allowing direct relationships between present filters. This is problematic for institutions, creating a variety of types to choose from with no real basis for comparison.

Both of the shape memory filters show advantages that cannot be found elsewhere. Nakagawa's filter was retrievable allowing short term protection against pulmonary emboli without the complications created by long term

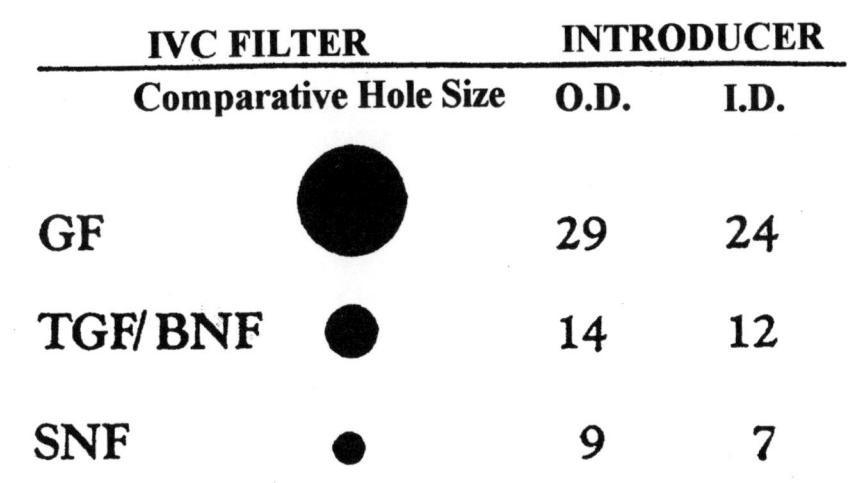

IVC FILTER		INTRODUCER	
Comparative Hole Size		O.D.	I.D.
GF	● (large)	29	24
TGF/BNF	● (medium)	14	12
SNF	• (small)	9	7

Fig 3.27 Schematic representation and comparison of puncture site holes and filter introduction of various modern IVC filters. GF = Greenfield Filter, TGF = Titanium Greenfield Filter; BNF = Bird's Nest Filter; SNF = Simon Nitinol Filter

placement. On the other hand, the Simon nitinol filter (SNF) uses a small 9F delivery catheter, approximately 3 mm in outside diameter. It can adjust to different sizes of lumen and its biocompatibility has been proved to be good. Furthermore, the SN filter has the smallest delivery system (Fig. 3.27) of any IVC filters presently available. This has made it possible for alternative placement routes such as antecubital vein and external jugular vein to be uniquely utilized. It does, however, have a problem with oversized lumen and it is advisable to restrict its use in caval diameters of less then 28 mm.

Whereas the Nakagawa filter is still in its initial stages, the SN filter is now in production by Nitinol Medical Technologies Incorporated. It is the opinion of most studies that the SN filter through its small size and associated advantages, such as ease of application, low risk of sequela, short overall operation time and possible safer introduction routes, has major benefits over its nearest rival, although further studies need to be carried out to allow direct comparison. The SN and Nakagawa filters represent a very useful step forward in IVC filter design and hence in the reduction of fatal recurrent pulmonary emboli.

References

(1) HOMANS, J. Deep quiet venous thrombosis in lower limb, *Surg Gynecol Obstet*, July 1944, **79**, 70–82.

(2) DOLAN, J. E., and ALPERT, J. S. Natural history of pulmonary embolism, *Progr Cardiovasc Dis*, 1975, **17**, 259–270.
(3) BERNSTEIN, E. F. The placement of venous interruption in the treatment of pulmonary embolism, *Pulmonary thromboembolism*, 1973, (Moser, K., Stein, M., eds) (Chicago year book medical publishers), 312–323.
(4) LINDENAUER, S. M. Prophylactic staple plication of the inferior vena cava, *Arch surg*, Dec 1969, **35**, 889–894.
(5) MORETZ, W. H., RHODE, C. M., and SHEPHARD, M. N. Prevention of pulmonary emboli by partial occlusion of the inferior vena cava, *Am Surg*, 1959, **25**, 617–626.
(6) DEWEESE, M. S., and HUNTER Jr, D. C. A vena cava filter for the prevention of pulmonary embolism, a 5 year clinical experience, *Arch Surg*, 1963, **86**, 852–868.
(7) OCHSNER, A., OCHSNER, J. L., and SANDERS, H. S. Prevention of pulmonary embolism by caval litigation, *Annual of Surgery*, 1970, **171**, 923–935.
(8) BLUMENBURG, R. M., and GELFAND, M. L. Long term follow up of vena clips., and umbrellas, *Am J of Surg*, 1977, **134**, 205–208.
(9) MOBIN-UDDIN, K., McLEAN, R., and JUDE, J. R. A new catheter technique for prevention of pulmonary embolism, *Am Surg*, 1969, **35**, 889–894.
(10) MOBIN-UDDIN, K., CALLARD, G. M., and BOBOKI, H. *et al*. Transfer caval interruption with umbrella filter, *New England J Med*, 1972, **286**, 55–58.
(11) STEIN, P. D., and SABBAH, H. N. Measured turbulence., and its effects on thrombus formation, *Circulation Research*, 1974, **35**, 608–614.
(12) GREENFIELD, L. J., McCURDY, J. R., BROWN, P. P., and ELKIN, R. C. A new intravascular filter permitting continued flow and resolution of emboli, *Surgery*, 1973, **73**, 599–606.
(13) GREENFIELD, L. J., PEYTON, R., CRUTE, S., and BARNES, R. Greenfield vena cava filter experience, *Arch Surg*, 1981, **116**, 1451–1456.
(14) GREENFIELD, L. J., ZOCCO, J., SCHROEDER, T. M., and ELKINS, R. C. Clinical experience with the Kimray–Greenfield vena caval filter, *Am Surg*, 1977, **185**, 692–698.
(15) GREENFIELD, L. J. Current indications for., and results of Greenfield filter placement, *J. Vasc Surg*, 1984, **1**, 502–504.
(16) MESSMER, J. M., and GREENFIELD, L. J. Greenfield caval filters: long term radiographic follow-up study, *Radiology*, 1985, **156**, 613–618.
(17) OSHNER PAIS, S., MIRVIS, S. E., and DE ORDIIS, D. F. Percutaneous insertion of the Kimray–Greenfield filter: technical considerations and problems, *Radiology*, 1987, **165**, 377–381.
(18) SIMON, M., KAPLAN, R., SALZMAN, E., and FREIMAN, D. A vena cava filter using thermal shape memory alloy, *Radiology*, 1977, **125**, 87–94.

(19) PALESTRANT, A. M., PRINCE, M., and SIMON, M. Comparative *in vitro* evaluation of the nitinol inferior vena cava filter, *Radiology*, 1982, **145**, 351–355.
(20) GRASSI, C. J. Inferior vena caval filters: analysis of five currently available devices, *AJR*, 1991, **156**, 813–821.
(21) KATSAMOURIS, A. A., WALTMAN, A. C., DELICHATSIOS, M. A., and ATHANASOULIS, C. A. Inferior vena cava filters: *in vitro* comparison of clot trapping., and flow dynamics, *Radiology*, 1988, **166**, 361–366.
(22) MOBIN-UDDIN, K., SMITH, P. E., and MARTINEZ, L. O. et al. A vena cava filter for the prevention of pulmonary embolus, *Surg Forum*, 1967, **18**, 209–211.
(23) LUND, G., RYSARY, J. A., and SALOMONWITZ, E. *et al*. A new vena cava filter for percutaneous placement and retrieval: experimental study, *Radiology*, 1984, **152**, 369–372.
(24) GUNTHER, R. W., SCHILD, H., FRIES, A., and STARKEL, S. Vena caval filter to prevent pulmonary embolism: experimental study, *Radiology*, 1985, **156**, 315–320.
(25) ROEHM Jr, J. O. F., GIANTURCO, C., BARTH, M. H., and WRIGHT, K. G. Percutaneous transcatheter filter for the inferior vena cava, *Radiology*, 1984, **150**, 255–257.
(26) ATHANASOULIS, C. A., ROBERTS, A. C., BROWN, K., GELLIER, S. C., WALTMAN, A. C., and ECKSTEIN, M. R. *The bird's nest versus the Kimray–Greenfield vena cava filter: randomized clinical study*, 1987, Presented at the 73rd Scientific Assembly and Annual Meeting of the Radiological Soc of North America.
(27) BURKE, P. E., MICHNA, B. A., HARVEY, C. F., CRUTE, S. L., SOBEL, M., and GREENFIELD, L. J. Experimental comparison of percutaneous vena caval devices: titanium Greenfield filter versus bird's nest filter, *J. Vasc Surg*, 1987, **6**, 66–70.
(28) SIMON, M., ATHANASOULIS, C. A., DUCKSOO, K., STEINBURG, F. L., and PORTER, D. H. *et al*. Simon nitinol inferior vena cava filter: initial clinical experience, *Radiology*, 1989, **172**, 99–103.
(29) TEITELBAUM, G. P., ORTEGA, H. V., and VINITSHIR, S. *et al*. Low artifact devices: MR imaging evaluation, *Radiology*, 1988, **168**, 713–719.
(30) DUCKSOO, K., SIEGEL, T. B., PORTER, D. H., and SIMON, M. Vena caval filter placement via the external jugular vein, *AJR*, 1990, **155**, 888–889.
(31) ROSE, S. C., and MILLER, F. J. Iced saline perfusion of Simon nitinol filters: improved control during transfemoral delivery: technical note, *Cardiovascular Intervent Radiol*, 1990, **13**, 111–112.
(32) DORFMAN, G. S. Percutaneous inferior vena cava filters, *Radiology*, 1990, **174**, 987–992.
(33) KASTAN, D. J., FORCIER, N. J., and KAHN, M. L. Simon nitinol

vena caval filter: preliminary observations., and suggested procedural modifications, *JVIR*, 1991, **2**, 123–124.
(34) DUCKSOO, K., SCHLAM, B. W., PORTER, D. H., and SIMON, M. Insertion of Simon nitinol caval filter: value of the ante-cusitar vein approach, *AJR*, 1991, **157**, 521–522.
(35) GRASSI, C. J., MATSUMOTO, A. H., and TEITELBAUM, G. P. Vena caval occlusion after Simon nitinol filter placement: identification with MR imaging in patients with malignancy, *JVIR*, 1992, **3**, 535–539.
(36) SCHAFER, A. I. The hypercoagulable states, *Ann Intern Med*, **102**, 814–828.
(37) KORBIN, C. D., REED, R. A., TAYLOR, F. C., PENTECOST, M. J., and TEITELBAUM, G. P. Comparison of filters in an oversized vena caval phantom: intracaval placement of a bird's nest filter versus biiliac placement of Greenfield, vena Tech-LGM,., and Simon nitinol filters, *JVIR*, 1992, **3**, 559–564.
(38) RICCO, J. B., CROCHET, D., and SEBILOTTE, P. *et al*. Percutaneous transvenous caval interruption with the 'LGM' filter: early results of a multicentre trial, *Ann Vasc Surg*, 1988, **3**, 242–247.
(39) McCOWAN, T. C., FERRIS, E. J., CARVER, D. K. L., and MOLPUS, W. M. Complications of the nitinol vena caval filter, *JVIR*, 1992, **3**, 401–408.
(40) HAWKINS, S. P., and AL-KUTOUBI, A. The Simon nitinol inferior vena cava filter: preliminary experience in the UK, *Clinical Radiol*, 1992, **46**, 378–380.
(41) LA PLANTE, J. S., CONTRACTOR, F. M., KIPROFF, P. M., and KHOURY, M. B. Migration of the Simon nitinol vena cava filter to the chest, *AJR*, 1993, **160**, 385–386.
(42) BECKER, D. M., PHILBRICK, J. T., and SELBY, B. Inferior vena cava filters: indications, safety, effectiveness, *Arch Intern Med*, 1992, **152**, 1985–1994.
(43) WINGERD, M., BERNHARD, V. M., MADISON, F., and TOWNE, J. B. Comparison of caval filters in management of venous thromboembolism, *Arch Surg*, 1978, **113**, 1264–1269.
(44) CIMOCHOWSKI, G. E., EVANS, R. H., and ZARINS, C. K. *et al*. Greenfield filter versus Mobin-Uddin umbrella, *J. Thorac Cardiovasc Surg*, 1980, 358–365.
(45) STEWART, J. R., PEYTON, J. W. R., CRUTE, S. L., and GREENFIELD, L. J. Clinical results of supravenal placement of the Greenfield filter, *Surgery*, 1982, **92**, 1–4.
(46) SCURR, J. H., JARRETT, P. E., and WASTELL, C. The treatment of recurrent pulmonary embolism: experience with the Kimray Greenfield vena cava filter, *Ann R Coll Surg Engl*, 1983, **65**, 233–234.
(47) SWANN, K. W., BLACK, P. M., and BAKER, M. F. Management of symptomatic deep venous thrombosis., and pulmonary embolism on a neurosurgical service, *J. Neurosurg*, 1986, **64**, 563–567.

(48) CARABASI III, R. A., MORITZ, M. J., and JARRELL, B. E. Complications encountered with the use of the Greenfield filter, *Am J. Surg*, 1987, **154**, 163–168.
(49) WHITNEY, B. A., and KERSTEIN, M. D. Thrombocytopenia and cancer: use of Kimray Greenfield filter to prevent thromboembolism, *South Med J.*, 1987, **80**, 1246–1248.
(50) GOLUEKE, P. J., GARRETT, W. V., and THOMPSON, J. E. *et al.* Interruption of vena cava by means of the Greenfield filter: expanding the indications, *Surgery*, 1988, **103**, 111–117.
(51) GREENFIELD, L. J., and MICHNA, B. A. Twelve year experience with the Greenfield vena caval filters, *Surgery*, 1988, **104**, 706–712.
(52) PAIN, S. O., TOBIN, K. D., AUSTIN, C. B., and QUERAL, L. Percutaneous insertion of the Greenfield inferior vena cava filter: experience with ninety six patients, *J. Vasc Surg*, 1988, **8**, 460–464.
(53) TODD, G. J., SANDERSON, J., and NOWYGROD, R. *et al.* Recent clinical experience with the vena cava filter, *Ann J. Surg*, 1988, **156**, 353–358.
(54) ROHRER, M. J., SCHEIDLER, M. G., WHEELER, H. B., and CUTLER, B. S. Extended indications for placement of an inferior vena cava filter, *J. Vasc Surg*, 1989, **10**, 44–50.
(55) HYE, R. J., MITCHELL, A. T., and DORY, C. E. *et al.* Analysis of the transition to percutaneous placement of Greenfield filters, *Arch Surg*, 1990, **125**, 1550–1553.
(56) KOLACHALAM, R. B., and JULIAN, T. B. Clinical presentation of thrombosed Greenfield filters, *Vasc Surg*, 1990, **9**, 666–670.
(57) FINK, J. A., and JONES, B. T. The Greenfield filter as the primary means of therapy in venous thromboembolic disease, *Surg Gynecal Obstet*, 1991, **172**, 253–256.
(58) CASTELLANI, L., NICAISE, H., and PIETRI, J. *et al.* Trenasvenous interruption of the inferior vena cava, *Int Angiol*, 1987, **6**, 299–306.
(59) FEBBE, F., DIETZEL, M., and KORTH, R. *et al.* Gunther vena caval filter: results of long-term follow up, *AJR*, 1988, **151**, 1031–1034.
(60) ROEHM Jr, J. O. F., JOHNSRUDE, I. S., BATH, M. H., and GIANTURCO, C. The Bird's Nest inferior vena cava filter: progress report, *Radiology*, 1988, **168**, 745–749.
(61) EPSTEIN, D. H., DARCY, M. D., and HUNTER, D. W. *et al.* Experience with the amplatz retrievable vena cava filter, *Radiology*, 1989, **172**, 105–110.
(62) GREENFIELD, L. J., CHO, K. J., and TAUSCHER, J. R. Limitations of percutaneous insertion of Greenfield filters, *J. Cardiovasc Surg*, 1990, **31**, 349–350.
(63) SCHNEIDER, P. A., GEISSBUHLER, P., and PIQUET, J. C. Follow up after partial interruption of the vena cava with the Gunther filter, *Cardiovasc Intervent Radiol*, 1990, **13**, 378–380.
(64) SIMON, M., RABKIN, D. J., and KLESKINKI, S. Comparative evalu-

ation of clinically available inferior vena cava filters with an in vitro physiologic simulation of the vena cava, *Radiology*, 1993, **189**, 769–774.
(65) NAKAGAWA, N., CRAGG, A. H., and SMITH, T. P. *et al.* A retrievable nitinol vena cava filter: experimental and initial clinical results, *JVIR*, 1994, **5**, 507–512.
(66) TADAVARTHY, S. M., CASTANEDA-ZUNGIA, W., and SALOMONOWITZ, E. *et al.* Kimray–Greenfield vena cava filter: percutaneous introduction, *Radiology*, 1984, **151**, 525–526.
(67) BOUAMEAUX, H., VELEBIT, V., BATTIKA, J., and SCHNEIDER, P. A. Z. *et al.* Recurrent, fatal pulmonary embolism despite inferior vena cava interruption with Greenfield filter, *Vasa*, 1987, **16**, 84–85.
(68) KANTOR, A., GLANZ, S., GORDON, D. H., and SCLAFANI, S. J. Percutaneous insertion of the Kimray–Greenfield filter, *AJR*, 1987, **149**, 1065–1066.

CHAPTER 4

The Use of the Ni–Ti SMA for Intravascular Applications

4.1 Intravascular endoprosthesis (IVEP)

4.1.1 Brief introduction to intravascular stents

The concept of intravascular stenting was first proposed in 1964 by Dotter and Judkins (**1**). In 1969 Dotter (**2**) published a further report which involved the intravascular (Fig. 4.1) positioning of stents in canines. Dotter implanted a variety of endoprostheses into the femoral and popliteal arteries of twenty-five dogs, via a percutaneous route with the entry site normally being the left carotid artery. The results gained are tabulated in Table 4.1

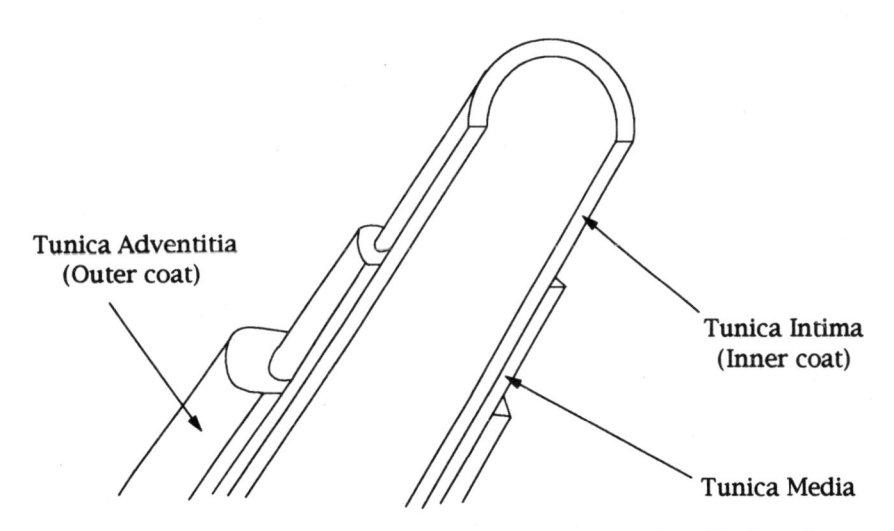

Fig 4.1   A diagrammatic representation of the layers constituting a blood vessel

Table 4.1 Constituent material and patency for the various intravascular endoprosthesis used by Dotter (2)

Type of endoprosthesis	Status: occluded/ patent	Time of occlusion/ extent of patency
Polyethylene	Occluded	24 hours
Polyamide	Occluded	24 hours
Silastic	Occluded	24 hours
Teflon	Occluded	24 hours
Silicon coated stainless steel coil	Occluded	24 hours
Uncoated stainless steel coil	Patent	2.5 years

Dotter discovered that coil prostheses showed good long-term patency. He commented on the advantages of coil prostheses, concluding that:

> An adoption of an open coil spring configuration has made possible long term patency. Advantages of this potentially useful approach include freedom from the trauma usually associated with surgical vascular reconstruction.

However, the apparent lack of interest in transluminal angioplasty, due to technical complications involved, meant that it was not until 1983 that any further development occurred. Since then a plethora of reports on various designs and materials have been published (3)–(10).

Intravascular (IV) stents were conceived to overcome the limitations of percutaneous transluminal angioplasty (PTA). PTA involves the insertion of a balloon tipped catheter in a peripheral part of the arteriovenous system to the stenotic region. Once in position the balloon is expanded and the lumen walls pushed out to reconstruct the original patent diameter. The balloon is then deflated and the catheter removed.

PTA has enabled movement away from more traumatic surgical remedies originally implemented, but suffers from a relatively high rate of restenosis in the dilated region. This restenosis occurs in approximately 30 percent of all PTA cases, even where anticoagulant schemes, antiplatelet therapy, anti inflammatory therapy and other systematic approaches have been implemented (11).

Acute dissection is another problem found in a small percentage of patients, approximately 5 percent in coronary patients and 2 percent in peripheral artery patients after PTA. This dissection originally resulted in the need for bypass surgery of the closure region. However, stents provide a possible non-surgical solution to this difficulty.

All vascular stents work on the same basic principle – they all hold up the walls of the artery by mechanical means, ensuring a clear pathway is produced. This stenting prevents elastic recoil exhibited by arterial walls, and may also trap plaque and other debris against the vessel walls, thus allowing a reduction

The Use of Ni–Ti SMA for Intravascular Applications

Table 4.2 Qualities pertaining to an ideal stent

Stable support	Flexibility
Good visibility	High expansion ratio
Ease of deployment	Flexibility
Neoendothelization and biologically inert	Reliable expandability
Thrombus resistant	Side branches maintained

in the possible thrombus generation at the dilation site. The ideal requirements for a stent are summarized in Table 4.2.

For ease, vascular stents can be separated into two basic groups according to their method of development:

(a) balloon expandable (Fig. 4.2)
(b) self-expandable.

The Ni–Ti thermally activated shape memory stent (Fig. 4.3) falls into the latter of these two groups, with its deployment being activated solely by the ambient temperature at placement site.

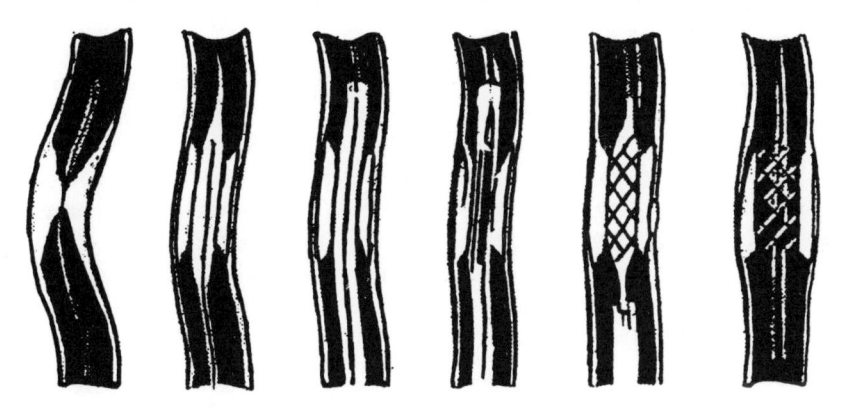

Fig 4.2 The Gianturco–Roubin balloon expandable coil stent (Katzen *et al.* (12))

Fig 4.3 Nitinol wire stent used by Dotter *et al.* (a) Compacted for catheter placement; (b) same coil after heat induced reversion (13). (Courtesy of RSnA.)

In recent years there have been many reports summarizing the progress gained so far in stent development **(14)–(17)**. The 1992 Katzen and Becker report entitled 'Intravascular Stents: Status Development and Clinical Application' was one such report. In their publication they discussed the future of vascular stents and commented that:

Although the significant benefit of stents in iliac transluminal anglioplasty has been realized, in the coronary, renal and femoral arteries, however, we are in the very early stages of this technology.

Even with the rapid development over the recent years, stents are still being enhanced and have not generally reached the commercial stage, with one exception, the Palmaz iliac balloon expandable stent **(18)–(21)**, that was recently approved by the American FDA.

Shape memory stents are one of the new generation of vascular devices and appear to fit most of the criteria required for an effective intravascular stent that involves a non-surgical placement.

4.1.2 Application of IVEPs

There are two main deployment techniques used to position nickel–titanium stents in the intravascular system.

(a) Cold saline technique – this involves a Ni–Ti wire whose TTR is around body temperature. The stent is transformed at a temperature well below its TTR and inserted into the delivery catheter where it is kept cool by the flow of cold saline around it. In reaching the desired PTA site it is pushed out of the catheter and regains its original shape as it warms to body temperature.

(b) Hot saline technique – this involves a Ni–Ti with a TTR just above body temperature. It is inserted using a catheter into the placement site, where warm saline is flushed over the stent and warms the SMA above its transition temperature, hence returning it to its original shape.

The use of Ni–Ti shape memory thermally activated stents was first reported in 1983 by two groups:

(a) Dotter *et al.* **(13)**
(b) Cragg *et al.* **(22)**.

The former of these two groups issued a report on the preliminary concepts for insertion using the hot saline technique of Ni–Ti coils into the femoral arteries of dogs (Fig. 4.3), and made no attempt to conclude any further than saying that:

Further technical refinements are the objects of our continuing interest.

Cragg *et al.* advanced on this. They inserted an Ni–Ti prosthesis (Fig. 4.4) into four canine aortae.

Fig 4.4 The loosely wound Ni–Ti prosthesis used by Cragg *et al*. (22). (Courtesy of RSnA.)

The group used an Ni–Ti alloy whose TTR was around body temperature, and straightened the wire in iced water. Then by using the cold saline technique they implanted the coil stent in the placement area. The group found that all stents showed good patency during the four week test period and concluded that Ni–Ti coil stents may:

> offer a simple, inexpensive alternative to surgery in numerous forms of cardiovascular disease.

In 1986, Sugita *et al*. (**23**) implanted twenty-two Ni–Ti coil ring stents (Table 4.3) into eleven dogs: twenty in the common iliac artery and two in the carotid artery.

Ten rings were surgically implanted into five dogs to evaluate the blood compatibility and the remaining twelve stents were placed via a catheter into a further six dogs. The catheter introduction procedure used by Sugita *et al*. was an adaption of the hot saline technique.

Once the coil (TTR = 40°C) was in position, the blood flow was halted by an expandable balloon at the end of the delivery catheter. 50°C saline solution was then flushed over the coil until the transition to original shape was complete. The balloon was then deflated and the catheter removed. The halting of blood was to allow the complete transformation of the coil stent under the influence of the hot saline solution, by preventing saline dispersion downstream.

From examination of the luminal surface of the Ni–Ti implant it could clearly be seen that a pseudoneointima had formed within one month of implantation. This lead Sugita *et al*. to conclude that the Ni–Ti alloy showed:

> excellent biocompatibility and long-term patency when implanted in canine arteries.

> This new technique shows promise for clinical application by preserving arterial patency and preventing restenosis following anglioplasty.

Table 4.3 Coil ring stent dimensions used by Sugita *et al*. (23)

Inner diameter	5 mm
Length	5 mm
Wire thickness	0.25–0.4 mm
Pitch	0–0.2 mm

Sugita *et al.* also suggested further applications of the SMA stent, discussing its possible usefulness for:

preserving the patency of the cerebral aqueduct, bladder neck, biliary tree and tracheobronchial tree.

As an ongoing experiment Sugita *et al.* left four dogs with implants for a further year.

In 1988, Kambic *et al.* (**24**) produced a follow up report on the four Sugita *et al.* canines. The group reported that all four stents showed 100 percent patency after the two year placement period, giving an overall patency rate of 91 percent (20/22 rings were patent up to one year). Kambic *et al.* concluded that, although their report only represented a small step, such stents may offer an alternative to conventional medical treatment.

In 1988 Sutton *et al.* (**25**) also reported on a two year study of Ni–Ti intravascular stent implants in dogs. The group used a coiled wire stent of the dimensions shown in Table 4.4

They implanted twelve of these endoprostheses, using the hot saline method, into the iliac ($n = 10$) and femoral ($n = 2$) arteries of six normal 25 kg dogs. The group examined both the patency of the lumen and the pathological effects of the implant in the placement region. Table 4.5 shows a summary of the degree of stenosis after implantation and it can be seen that all stents were patent after 1 and 2 years, with only a small amount of stenosis occurring in a few prostheses.

The histological study showed that a thin neointima layer formed soon after implantation without the presence of inflammatory cells (Table 4.6).

At two months a smooth muscle build up had appeared at both proximal and distal ends of the prosthesis to form a smooth surface between implant and natural vessel wall, and by the end of the two year study, a stable endothelial-like cell layer could be seen covering the inner surface of the implant.

Sutton *et al.* concluded that:

The Ni–Ti intravascular endoprosthesis showed satisfactory delivery, excellent patency and biocompatibility, with rapid endothelialization and stable neointimal thickness over two years.

In another report in the same year, Oku *et al.* (**26**) reported on an *in vitro* study designed to evaluate the implantation technique and reliability of transition at various temperatures of a miniaturized Ni–Ti intravascular endoprosthesis (IVEP) (Table 4.7).

Oku *et al.* set up a mock circulatory loop (MCL) to simulate *in vivo* conditions, using a pulsatile flow of 100 ml/min of water at body temperature. The coil design IVEP was implanted into a test section of the MCL. Various amounts and temperature of hot saline were injected via a catheter to enable the IVEP to convert.

Two types of catheter design were used: Model 68-1-3.0 had a single 30 mm long slot compared to Model 87-1-3.0, which had two parallel 5 mm slots. Both

Table 4.4 Ni–Ti dimensions used by Sutton et al. (25)

Length	5 mm
Inner diameter	5 mm
Wire thickness	0.25 mm

Table 4.5 Summary of implantation patency for the Ni–Ti implants used by Sutton et al. (25)

Prostheses No.	Implantation site	Degree of stenosis (%)
1 year implants		
1	LF	50†
2	R1	0
3	L1	0
4	R1	0
5	L1	10
6	R1	10
7	L1	0
8	R1	20
2 year implants		
9	LF	20
10	R1	0
11	L1	0
12	R1	0

Note: LF = left femoral artery; R1 = Right external iliac artery; L1 = Left external artery.
*Mean stenosis for 1 year implants 11 ± 17%; mean stenosis for 2 years implants 5 + 10%.
†Prosthesis *1 was occluded at 1 month and recanalized at 6 months.

Table 4.6 Mean neointimal thickness after implantation in iliac and femoral arteries by Sutton et al. (25)

Implant duration and location	Neointimal thickness (mean ± SD in μm)			
	Proximal	Middle	Distal	Entire implant
1 year				
Left	171.3 ± 95.6	333.4 ± 123.3	461.0 ± 98.1	311.2 ± 85.9
Right	83.0 ± 5.2	118.3 ± 12.5	115.5 ± 29.0	106.5 ± 5.1
Left & right combined	136.0 ± 83.2	247.3 ± 146.7	322.8 ± 202.1	229.3 ± 127.6
2 years				
Left	180.3 ± 165.0	398.2 ± 450.7	316.2 ± 327.2	298.6 ± 309.7
Right	118.4 ± 118.0	164.5 ± 184.1	188.9 ± 185.5	148.0 ± 149.4
Left & right combined	149.3 ± 122.4	281.4 ± 311.8	252.5 ± 229.2	223.3 ± 216.7

Table 4.7 Dimensions of Ni–Ti stent by Oku et al. (26)

Length	5 mm
Inner diameter	2.5 mm
Wire dimensions	0.15 mm × 0.38 mm

types involved the inflation of a balloon to temporarily halt transluminal flow, and thus raise local temperature by preventing 'heating' saline to disperse downstream.

From the results gained it could be seen that there were some inconsistencies. Oku et al. explained these by highlighting the minor variations between the two catheter designs and between 'identical' injections (T_i and Vol_i) for each catheter, and also by noting that, as all the IVEPs were hand wound onto the catheter, the amount of strain on each may vary.

Oku et al. concluded that consistent activation could be achieved, however, by 5 ml of 70°C normal saline which only increased the local temperature (T_p) to 55° for less than 1 second, and adverse effects to this warming had not been detected in previous in vivo studies.

In 1989, Tominaga et al. (**27**) published a report on the effects of gaps between wire pitches of an Ni–Ti thermally activated intravascular stent on restenosis and side branch patency. In this study forty-two stents were transluminally implanted into ten atherosclerotic and eleven normal New Zealand white rabbits, with two stents placed in each rabbit.

The rabbits underwent balloon angioplasty of the infrarenal abdominal aorta and two stents, one with 0.33 mm gaps and one without gaps, were placed using the hot saline technique (10 ml of 75°C saline) into the treated area. The proximal and distal positions of the two types of stent were alternated and the effect on luminal patency for both atherosclerotic and normal rabbits obtained after set intervals of time.

Rapid occlusion within four weeks occurred in one of the normal rabbits and in one of the atherosclerotic animals. Also one of the normal rabbits showed severe stenosis after four weeks and complete occlusion after eight weeks in its ungapped stent.

However, Tominaga et al. observed that:

> Surface geometry of the IVEP influences the speed of endothelialization and neointimal thickness. The placement of gaps between wire pitches apparently promotes rapid endothelial coverage of the stent, which is expected to help prevent restenosis.

Tominaga et al. also reported on the patency of major side branches around the implantation area, and noted that areas with IVEPs containing gaps showed a higher patency rate than the corresponding areas of ungapped stents.

Tominaga concluded that:

> These results indicate that placing gaps between wire pitches was beneficial in reducing neointimal thickness and restenosis and preventing side branch obstruction.

In a very similar report published in 1992, Tominaga et al. (**28**) reported on the effect of gapped and ungapped stents in ten normal rabbits. The stent's dimensions were the same as described in the previous report and the

Table 4.8 DSA Measurements of maximum stenosis vessels by Tominaga 1992 (millimetres, mean ± SD) (28)

Implant duration	Stents with gaps	Stents without gaps	Significance
4 weeks ($n = 10$)	2.97 ± 0.22	2.87 ± 0.23	Not significant
12 weeks ($n = 9$)	2.94 ± 0.16	2.72 ± 0.22	$p < 0.005$
16 weeks ($n = 9$)	2.92 ± 0.18	2.62 ± 0.23	$p < 0.005$
20 weeks ($n = 8$)	2.83 ± 0.12	2.51 ± 0.15	$p < 0.001$
24 weeks ($n = 8$)	2.77 ± 0.19	2.37 ± 0.30	$p < 0.01$

implantation procedure was also very similar to that used by Tominaga et al. (27). Extensive analysis of the rabbits was performed involving digital subtraction angiograms (DSA) and histopathological examinations. The results gained by Tominaga et al. are shown in Table 4.8 and Fig. 4.5.

The group reported that only one ungapped stent showed total occlusion during the test period, and this formed after twelve weeks. Tominaga et al.'s findings were similar to the 1989 report and the group concluded in a similar manner:

These results indicated that placing gaps between wire pitches was

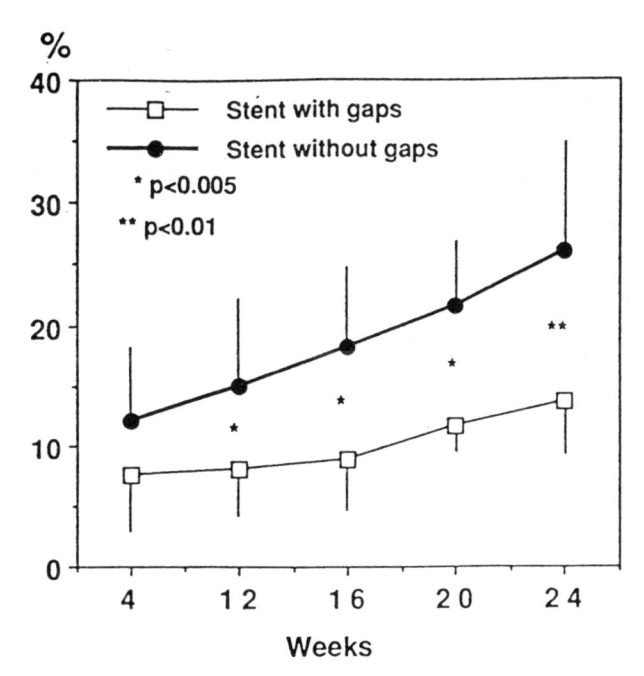

Fig 4.5 Changes in percent stenosis. Starting at 12 weeks there were statistically significant differences between the two Tominaga groups (1992). Percent stenosis was calculated as (the stent internal diameter − the narrowest internal diameter)/the stent internal diameter (28)

beneficial in reducing neointimal thickness and stenosis rate as well as in preventing side branch obstruction after balloon angioplasty.

The group further concluded that:

Further study using the atherosclerotic model is necessary, because this study suggests that improved design with the introduction of gaps may be helpful to prevent restenosis after PTA.

In 1993 Tominaga et al. (**29**) followed up their initial investigations into the effect of stent design on restenosis by reporting on a further study of nineteen rabbits. In this study Tominaga et al. assessed the effects of stent geometry and reaction to high levels of cholesterol within the blood serum of atherosclerotic animal subjects.

The group split the animals into two study classes with both classes on a pre-operatively high cholesterol diet. Class I ($n = 9$) was returned to normal food intake after stent implant, whereas Class II ($n = 10$) remained on the high cholesterol diet.

Tominaga et al. implanted one of their NiTi SMA stents (Table 4.9) into each animal using a warm saline technique and assessed the affects of geometry and cholesterol levels after set periods of 4, 8, 12, 16, 20 and 24 weeks. The mean results are shown in Fig. 4.6.

Tominaga et al. commented that:

There was a statistically significant difference in the percent stenosis within the closed stents between groups I and II.

Favourable trends of lowering cholesterol were observed in a comparison of the percent stenosis within the open stents between the two groups.

The group concluded that:

Both stent design and serum cholesterol are important factors for stenosis after stent implantation.

In 1993 a group headed by Eigler (**30**) reported on the use of nitinol SMA stents as temporary stents that are removable hours or days after implantation.

The idea of a temporary stent is to provide most of the benefits of a permanent stent and to try to mitigate the problems of thrombosis and prolonged anticoagulant therapy after implantation.

Table 4.9 Dimensions of stent used by Tominaga et al. (1993) (29)

Feature	Closed stent	Open stent
Wire diameter (mm)	0.33	0.33
Inner diameter (mm)	3.2	3.2
Prosthesis outer diameter (mm)	3.5	3.5
Prosthesis length (mm)	15	15
Prosthesis wire pitch (mm)	–	0.33

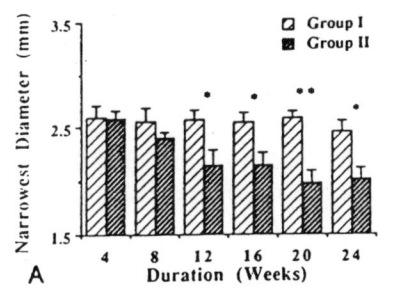

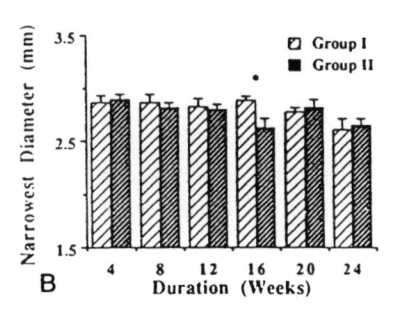

Fig 4.6 Changes in narrowest diameter of arterial segments as observed by Tominaga *et al.* (29) from (A): closed stent; (B): open stent

Eigler *et al.* implanted seventy-eight stents (transition temperature, 55°C) into the left descending anterior coronary artery ($n = 37$) and the left circumflex coronary artery ($n = 41$) of twenty-eight mongrel dogs. The collapsed stent was positioned at the placement site and expanded using a modification of the balloon angioplasty technique. The prostheses were left for 1–6 weeks (Table 4.10) and then removed.

Catheter delivery of 3–5 ml of 75–80° lactated Ringer's solution around the placement site raised the ambient temperature to above 55°C and caused the stent to return to its memorized, collapsed, form, which gripped the recovery/Ringer injection catheter so that both catheter and stent could be retrieved.

Eigler *et al.* found that all seventy stents implanted for up to one week were successfully removed. However, stents in place for 6 weeks were not able to be retrieved due to the proliferation of neointimal tissue that had formed over the endoprosthesis. The group commented that:

> the ease of stent recovery was the most noticeable characteristic of the device, and that there had been no instances of device embolization during placement or recovery.

Table 4.10 Location and duration of stent implantation applied by Eigler *et al.* (30)

No. of dogs ($n = 28$)	Stents implanted		Duration (hours)			
	LAD ($n = 37$)	LCx ($n = 41$)	0.5	1.5–5	24	72
2	4	3	7	–	–	–
10	17	22	23	16	–	–
4	4	4	–	–	8	–
4	4	4	–	–	–	8
4	4	4			1 wk	
2	2	2			6 wk	
2	2	2			6 mo	

LAD = Left anterior descending coronary artery; LCx = left circumflex coronary artery.

Eigler *et al.* noted that the clinical value of temporary stents had not been demonstrated, but hypothesized that a removable stent:

> may have a role in treatment of flow compromising intimal dissection because prolonged balloon inflation for between 20 minutes and 15 hours has been reported to stabilize lumen-compromising dissections. Potentially, intimal 'healing' for several hours to a few days could stabilize the dissection yet still allow the stent to be removed before the time of peak incidence of subacute thrombosis (days 4–10).

Eigler *et al.* summarized their findings by saying they had:

> demonstrated the feasibility of a new method for temporary stenting of the coronary arteries that uses the special thermoelastic properties of nitinol to permit satisfactory deployment, balloon expansion and recovery of the device in normal canine coronary arteries.

In another report involving the stenting of coronary arteries, Grenadier *et al.* (**31**) investigated the affects of nitinol shape memory alloy endoprosthesis (transition temperature <37°C) in twenty-three mongrel dogs. Grenadier *et al.* implemented neither a hot nor cold saline technique, but instead, the group constrained the prosthesis mechanically until at the placement site. Once in position the constraints were removed, and the SMA allowed to recover its shape.

Grenadier *et al.* lost five dogs to the follow up. However, no losses were due to actual prosthesis failure or thrombosis/embolization at the placement site.

The group compared the tissue proliferation and inflammatory response to the nitinol stent to other common types of stents and commented that the NiTi endoprosthesis showed:

> acute and long term patency that is no different from other stents.

Grenadier *et al.* concluded that, due to

> its low metal content and very high longitudinal flexibility and radial conformability, this stent may have impact clinical application.

Since the mid 1980s the iliac arteries have been the site of endoprosthesis assessment. In 1994 Hausegger *et al.* (**32**) produced a paper describing their experience with a NiTi SMA stent (Table 4.11) in fourteen patients with arteriosclerosis (ten men and four women).

The group implanted the stent using a cold saline technique, and recorded the transstendic pressure gradient before and after the implantation. The patient data are shown in Table 4.12.

Table 4.11 Dimensions of stent used by Hausegger *et al.* (32)

Length	31 or 62 mm
Inside diameter	8 or 10 mm
Wire diameter	0.27 mm

Table 4.12 Patient characteristics and initial and late results after stent implantations as observed by Hausegger et al. (32)

Patient no.	Stenosis location	Extent of stenosis (percentage of diameter)		Indication for* stent placement	Stage†		Ankle-brachial index			Follow up (months)	Adjunctive treatment
		Before treatment	After treatment		Before treatment	At last follow up	Before treatment	After treatment	At last follow up		
1	EIA	90	0	Dissection	IIb	I	0.6	0.83	0.88	15	None
2	EIA	80	0	Restenosis	IIb	I	0.6	0.77	1.11	24	Femoral PTA
3	CIA	100	50	Occlusion	IIb	I	0.0	0.87	1.38	24	None
4	CIA	100	0	Occlusion	IV	NA	0.43	0.35	NA	Lost	Femoral bypass
5	EIA	90	30	Gradient	IIb	I	0.29	0.69	0.58	15	None
6	EIA	80	0	Gradient	IIb	I	0.0	0.75	0.93	12	Femoral PTLA
7	CIA	70	0	Gradient	III	I	0.0	0.3	0.76	15	None
8	CIA	80	0	Gradient	IIb	I	0.7	1.0	0.93	6	None
9	CIA	100	0	Occlusion	III	I	0.0	0.8	1.05	12	None
10	CIA	100	0	Occlusion	IIb	I	0.8	1.3	1.25	6	None
11	EIA	70	0	Restenosis	IIb	IIa	1.0	1.1	0.88	14	None
12	CIA	95	0	Dissection	IIb	I	0.45	1.0	1.3	6	Femoral bypass
13	EIA	90	0	Gradient	IIb	IIa	0.41	0.6	0.78	11	None
14	EIA	100	0	Occlusion	IIb	I	0.45	1.0	1.0	6	None

Note: CIA = common iliac artery; EIA = external iliac artery; NA = not available; PTA = percutaneous transluminal angioplasty; PTLA = percutaneous transluminal laser angioplasty.
*Gradient = post-PTA transstenotic mean pressure gradient of > 10 mm Hg
†Stages were determined by using the Fontaine classification system: I = free of symptoms; IIa = mild claudication (walking distance >250 m); IIb = advanced claudication (walking distance <250 m); III = ischemic at rest; IV = gangrene.

One patient was lost to follow up (patient 4 in Table 4.12) and in two further patients (patients 3 and 5 in Table 4.12) the radial force exerted by the stent was not great enough to provide complete vessel recovery. All other patients showed excellent vascular restoration and the group observed that no complications, recurrent stenosis or stent occlusions occurred. Hausegger *et al.* concluded that:

> the good radiopacity and the minimal shortening of the Cragg stent make exact placement easy. Because of the fairly good longitudinal flexibility, effective treatment of tortuous vessels is possible.
>
> The thermal recovery properties of nitinol could be a major advantage in this field.
>
> Furthermore, since nitinol is not a ferromagnetic alloy, follow up studies can be performed non-invasively with magnetic resonance angiography. Therefore the future 'ideal' stent might be made of nitinol.

In a further report in 1994 Beyar *et al.* (**33**) reviewed the initial cardiovascular experience of a nitinol SMA stent, called the 'Instent'. Beyar's Instent was very similar in design to the Grenadier stent and the group used an almost identical method of mechanically constrained placement.

The group implanted the stent into twenty-three dogs and the peripheral arteries of seven human patients. The results are shown in Tables 4.13 and 4.14.

Table 4.13 Angiographic dimensions (D) and intimal thickness (IT) at follow up for the twenty-seven sheep implemented by Beyar *et al.* (33)

Duration	1–2 wks (n = 2)	1 mo (n = 4)	3 mo (n = 4)	6 mo (n = 4)	12 mo (n = 2)
D (mm) control	2.6	2.9	2.6	2.6	3.0
D (mm) follow up	2.6	2.7	2.6	2.7	3.0
IT (μ)	30 ± 10	141 ± 105	227 ± 104	211 ± 99	170 ± 42

Table 4.14 Description of site, indications and follow up results gained from stents implanted into the peripheral arteries in Beyar *et al.*'s patients (33)

Case	Site	D (mm)	Indication	Result	Follow up 3–6 mo
1	Popliteal	5	Restenosis	Fav†	Patent
2	Femoral	6	Dissection	Fav	Restenosis within and outside stent
3	Femoral	6	Dissection	Fav	Patent-restenosis outside stent
4	Femoral	6	Dissection	Fav	Patent
5	Graft	7	Restenosis	Fav	Patent
6	Iliac	8	Dissection	Fav	Patent
7	Femoral	6	Dissection	Fav	Patent

†Favourable

Beyar *et al.* discussed the ability of this 'constrained' stent as a temporary or removable endoprosthesis and concluded that although the stents had so far shown

> good results with an acceptable restenosis rate. The self expandable mechanism has been shown to be practical in patients and good tacking of arterial dissections was demonstrated.

The group went on to say that the prosthesis required

> further larger scale studies, both in the experimental laboratory and in patients to precisely determine the advantages and limitations of the current stent as compared to other stents.

In 1993 Cragg *et al.* (**34**) published a paper comparing their nitinol SMA intravascular stent to the commercially available self-expanding Wallstent™. The group delivered forty-four 'Cragg' stents, Table 4.15) into twenty-two healthy adult sheep weighing between 50 and 70 kg.

Three sheep were lost to the follow up. With the remaining nineteen sacrificed at one month ($n = 2$), three months ($n = 2$), six months ($n = 8$) or left for long-term observation ($n = 7$).

Of the twenty-eight stents accessible to follow up, none of them showed any adverse tissue reaction around the stent and any inflammation was mild. The group found that the radial force exerted by the SMA stent was favourable in comparison with the commercially available Wallstent and observed only an average of 11 percent restenosis at six months (Fig. 4.7).

Cragg *et al.* concluded that:

> The animals in our study showed no local or systemic toxicity related to the implants.
>
> Our results indicate that our stent can be reliably and safely deployed in the vascular system.

Although a lot of investigations have been made into nickel titanium stents, they still have not been clinically applied. A lot of factors affect stent efficiency, such as diameter, length, material, surface geometry, radial compliance, and so on, and the aggravated damage created by percutaneous transluminal balloon angioplasty to the endothelial layer and hence the encouragement of thrombi formation. The need to investigate all these factors is vital. However, a lot of work has been done, and Ni–Ti thermally activated IVEPs appear to show great promise in the field of intravascular stenting.

Table 4.15 Dimensions of stent used by Cragg *et al.* (34)

Length	60 mm
Inside diameter	10 mm
Wire diameter	0.28 mm

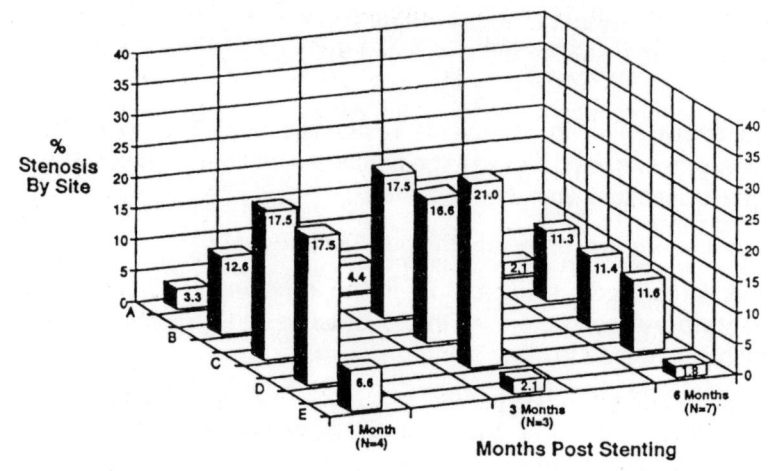

Fig 4.7 Results gained by Cragg et al. (34). (Courtesy of RSnA.)

4.2 Other intravascular devices

The occlusion of intravascular vessels can be achieved using many techniques. However, with large vessels where the flow is non laminar and/or rapid, the rate of failure or complication is high. In these cases a theoretical prosthesis could be inserted into the lumen and, by expansion and attachment to vessel wall, achieve occlusion with none of these functional failures.

In 1992 Ernest Marni (**35**) published a report on the use of shape memory fixture needles in the occlusion of large arteries. Marni's prototype is shown more clearly in Fig. 4.8.

The prosthesis consisted of a catheter introduced polyurethane balloon with Ni–Ti alloy needles. On implantation and expansion of balloon at site, hot saline (60°C) was injected into the balloon and the needles returned to their memorized shape, thus attaching the prosthesis to the vessel wall and preventing migration downstream. Through very basic testing, Marni concluded that his prototype:

> illustrated employing two different materials is practical for occluding large vessels. This initial experiment demonstrated the conceptual validity of the principle and the possibility of practical application. Further research and experimental use in laboratory animals are necessary to determine the optimal size and shape of needles.

Whereas Marni used a catheter introduced prosthesis with SMA needles, Rosenthal et al. (**36**) reported in 1992 on the use of an electronically steerable Ni–Ti catheter in *in situ* saphenous vein bypass. The mechanics of an *in situ* saphenous bypass involves the placement of prostheses to occlude the venous side branches, and hence encouraging the embolization of the pathways.

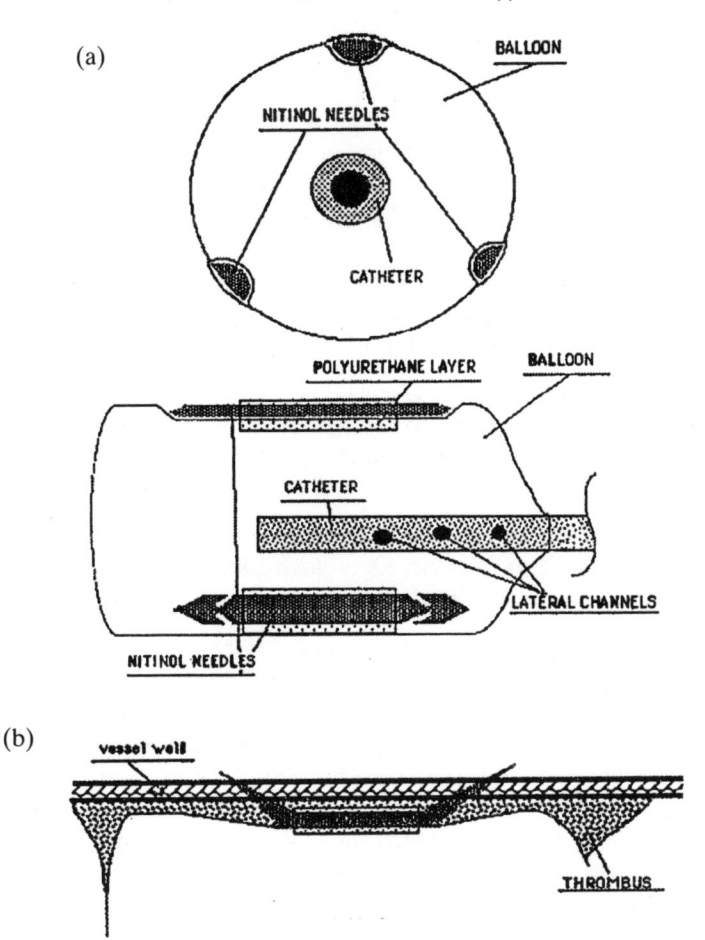

Fig 4.8 Prosthesis used by Marni in 1992 (35). (a) Transverse and longitudinal section of balloon and needles before activation; (b) transition completed after injection of 60°C saline needles in final position and prosthesis securely attached to vessel wall

Rosenthal *et al.* performed forty-six endovascular *in situ* bypasses of either femoral distal popliteal or tibial veins. The group used a shape memory catheter to place prostheses in the forty-four patients. This catheter was steered by the application of steering microfilaments. Rosenthal *et al.* found that 82 percent of the side branched were totally occluded and 18 percent partially occluded by the end of the study, and although the group gave no technical information on the catheter, they concluded that:

This initial clinical study demonstrated that an electronically steerable shape memory metal alloy catheter can be used to occlude venous tributaries safely and effectively from within the saphenous vein without affecting early bypass patency.

The group also discussed the need for further refinement of the catheter design but gave an optimistic conclusion on initial testing.

In another report headed by Rosenthal (37) a group evaluated the feasibility of endoluminal angioscopic-assisted occlusion of venous tributaries using a NiTi SMA catheter.

The group implemented a microfilament heating technique similar to one previously mentioned (36) and performed various investigations on either *in vitro, in vivo* dogs or in amputated limbs to assess the NiTi catheter. Rosenthal *et al*. concluded that although

> further improvement was necessary, their preclinical studies demonstrated that under angioscopic and fluoroscopic control the steerable nitinol catheter can be used to achieve occlusion of venous tributaries.

In 1994 yet another group headed by Rosenthal (38) reported on a further fifty patients that had undergone 53 endovascular *in situ* saphenous vein (EISV) bypasses. Rosenthal *et al*. discussed the advancements in catheter design and procedural technique compared to the 1992 report (36) and found the hospital length of stay (LOS) to be greatly reduced in comparison to traditional ('classic') techniques. Rosenthal *et al*. concluded:

> If EISV bypass long term patency rates remain similar to classic *in situ* bypass patency rates, the additional benefits of decreased hospital LOS, reduced wound-related complications, shortened recuperation and therefore increased health care savings give the endovascular concept strong consideration as a possible future operation for infraguinal saphenous vein *in situ* bypass.

The occlusion of arteries by artificial means is sometimes vital for the welfare of patients. Rosenthal has reported on the bypassing of the saphenous vein (36)–(38), and in 1994 Huang *et al*. (39) produced a report investigating the ability of a NiTi SMA endoprosthesis to occlude the external iliac arteries in four pigs and two dogs.

Huang *et al*. used a conic prosthesis made from 0.35 mm diameter wire and with the larger proximal lumen of 4 mm in the dogs or 4.5 mm in the pigs. The two dogs and four pigs were sacrificed at six and eight months, respectively, and the group found that in both cases occlusion of the external iliac had proceeded satisfactorily. The group mentioned that

> neither inflammatory cells infiltration nor foreign body giant cell reaction was found in the vessel wall.

Huang *et al*. concluded that

> Conic nitinol alloy stents may completely occlude arterial blood flow.

4.3 Conclusion

From their advent into intravascular applications in the late 1960s, much research time has been allocated for the advancement of the Ni–Ti SMA alloy within this highly sensitive field.

Much of the research undertaken has involved the application of SMA endoprostheses, or stents as they are more commonly known. These stents have shown good term patency within canines (23), and have not been observed to show any detrimental morphological or histological effects on the surrounding tissue of the lumen.

Intravascular endoprosthetic techniques have been developed to overcome the problems caused by vessel stenosis and the relative amount of success of percutaneous transluminal angioplasty on preventing restenosis.

The small angioplastic catheter insertion technique made possible by the SMA's unique characteristics make it less likely for substantial turbulence or damage to occur whilst placement is being achieved. So large advantages over other possible endoprosthetic methods can be seen.

Ni–Ti SMA coil stents geometrically designed with gaps to maintain patency have made large steps towards the 'ideal' stent. Although not clinically applied as yet, they may prove to be invaluable in future treatment of intravascular stenosis and prevention of restenosis.

Although endoprosthetics is not the only region of SMA intravascular investigation, it does possess the largest background of information. In more recent years, studies into the optimization of various other intravascular techniques by using the SMA have been researched. All have shown encouraging results and have gone a long way to proving the vast potential for the application of the Ni–Ti SMA within the intravascular field.

It is believed by the author of this book that the Ni–Ti SMA shows outstanding characteristics and the possibilities for the solution of various methodological and physiological problems, although requiring further investigation, look very encouraging indeed.

References

(1) DOTTER, C. T., and JUDKINS, M. P. Transluminal treatment of arteriosclerotic obstruction, *Circulation*, 1964, **30**, 654.

(2) DOTTER, C. T. Transluminally placed coil-spring end arterial tube grafts, long term patency in canine popliteal artery: Investigation, *Radiology*, 1969, **4**, 329.

(3) MAASS, D., ZOLLIKOFER, CHL., LARGIADER, F., and SENNING, A. Radiological follow-up of transluminally inserted vascular endoprostheses: An experimental study using expanding spirals, *Radiology*, 1984, **150**, 45.
(4) WRIGHT, K. C., WALLACE, S., CHARNSANGAVEJ, C., CARRASCO, C. H., and GIANTURCO, C. Percutaneous endovascular stents: an experimental evaluation, *Radiology*, 1985, **156**, 69.
(5) CHARNSANGAVEJ, C., WALLACE, S., WRIGHT, K. C., CARRASCO, C. H., and GIANTURCO, C. Endovascular stent for use in aortic dissection: an *in vitro* experiment, *Radiology*, 1985, **157**, 323.
(6) SIGWART, U., PUEL, J., MIRKOWITCH, V., JOFFRE, F., and KAPPENBERGER, L. Intravascular stents to prevent occlusion and restenosis after transluminal anglioplasty, *New Eng J. Med*, 1987, **316**, 701.
(7) PALMAZ, J. C., SIBBITT, R. R., REUTER, S. R., TIO, F. O., and RICE, W. J. Expandable intraluminal graft: preliminary study, *Radiology*, 1985, **156**, 73.
(8) SCHATZ, R. A., and PALMAZ, J. C. Balloon expandable intravascular stents for anglioplasty, *Cardio*, 1987, **4**, 27.
(9) ROLLINS, N, WRIGHT, K. C., CHARNSANGAVEJ, C., WALLACE, S., and GIANTURCO, C. Self expanding metallic stents: preliminary evaluation in an atherosclerotic model, *Radiology*, 1987, **163**, 739–742.
(10) DeJAEGERE, P. P., STRAUSS, B. H., VAN DER GIESSEN, W. J., FEYTER, P. J., and SERRUYS, P. W. Immediate changes in stenosis geometry following stent implantation: comparison between a self expanding and balloon expanding stent, *J. Invent Cardio*, 1992, **5**, 71–78.
(11) KARAS, S. P., SANTOIAN, E. C., and GRAVANIS, M. B. Restenosis following coronary angioplasty, *Clinical Cardiology*, 1991, **14(10)**, 791–801.
(12) KATZAN, B. T., and BECKER, G. J. Intravascular stents, *Endo Vasc Surg*, 1992, **72(4)**, 941–957.
(13) DOTTER, C. T., BUSCHMANN, R. W., MCKINNEY, M. K., and ROSCH, J. Transluminal expandable nitinol coil stent grafting: preliminary report, *Radiology*, 1983, **147**, 259–260.
(14) SCHATZ, R. A. A view on vascular stents, *Circulation*, 1989, **79(2)**, 460–462.
(15) KING III, S. B. Vascular stents and atherosclerosis, *Circulation*, 1989, **79(2)**, 460–462.
(16) PORTER, J., ALISAN, A., MULCALY, D., and SIGWART, V. Coronary stents, *Brit J. Hosp Med*, 1992, **47(6)**, 411–419.
(17) KATZAN, B. T., and BECKER, G. J. Intravascular stents, *Endo Vasc Surg*, 1992, **72(4)**, 941–957.

(18) PALMAZ, J. C., GARCIA, O., and SCHATZ, R. A. *et al.* Placement of balloon expandable stents in iliac arteries: first 171 patients, *Radiology*, 1990, **174**, 969–975.
(19) PALMAZ, J. C., TIO, F. O., and SCHATZ, R. A. *et al.* Early endothelization of balloon expandable stents: experimental observation, *J. Intervent Radiol*, 1988, **3**, 119–124.
(20) PALMAZ, J. C., RICHTER, G. M., and NOELDGE, G. *et al.* Intraluminal stents in atherosclerotic artery stenosis: preliminary report on multicentre trial, *Radiology*, 1988, **168**, 727–731.
(21) PALMAZ, J. C., LABORDE, J. C., and RIVERA, F. J. *et al.* Three year experience with iliac stents (Abstract), *Radiology*, 1991, **181**(p), 244.
(22) CRAGG, A., LUND, G., RYSAVY, J., and CASTANEDA, F. *et al.* Non surgical placement of arterial endoprosthesis: A new technique using nitinol wire, *Radiology*, 1983, **147**, 261–263.
(23) SUGITA, T., SHIMOMITSU, T., OKU, T., and MURABAYASHI, S. *et al.* Non surgical implantation of a vascular ring prosthesis using thermal shape memory Ni–Ti alloy (nitinol wire), *Trans ASAIO*, 1986, **32**, 30–34.
(24) KAMBIC, H., SUTTON, C., OKU, T., and SUGITA, Y. *et al.* Biological performance of Ni–Ti shape memory alloy vascular ring prosthesis: a two year study, *Int J. Arti Organs*, 1988, **11(6)**, 487–492.
(25) SUTTON, C. S., OKU, T., HARASKI, H., and KAMBIC, H. E. *et al.* Titanium–nickel intravascular endoprosthesis: a 2 year study in dogs, *AJR*, 1988, **151**, 597–601.
(26) OKU, T., SUTTON, C. S., KAMBIC, H. E., HARASKI, H., and NOSE, T. A titanium–nickel alloy intravascular endoprosthesis: *in vitro* studies, *Trans ASAIO*, 1988, **34**, 399–403.
(27) TOMINAGA, R., EMOTO, H., KAMBIC, H. E., and HARASKI, H. *et al.* Intravascular endoprostheses: effect of surface geometry on restenosis and side branch patency, *Trans ASAIO*, 1989, **35**, 376–378.
(28) TOMINAGA, R., KAMBIC, H. E., EMOTO, H., and HARASKI, H. *et al.* Effects of design geometry of intravascular endoprostheses on stenosis rate in normal rabbits, *Am Heart J.*, 1992, **123(1)**, 21–28.
(29) TOMINAGA, R., HARASKI, H., and SUTTON, C. *et al.* Effects of stent design and serum cholesterol level on the restenosis rate in arteriosclerotic rabbits, *Am Heart J.*, 1993, **126(5)**, 1049–1058.
(30) EIGLER, N. L., KHORSANDI, M. J., and FORRESTER, J. S. *et al.* Implantation and recovery of temporary metallic stents in canines coronary arteries, *JACC*, 1993, **22(4)**, 1207–1213.
(31) GRENADIER, E., SHOFTI, R., and BEYAR, M. *et al.* Self-expandable and highly flexible nitinol stent: immediate and long term results in dogs, *Am Heart J.*, 1994, **128**, 870–878.
(32) HAUSEGGER, K. A., CRAGG, A. H., and LAMMER, J. *et al.* Iliac artery stent placement: clinical experience with a nitinol stent, *Radiology*, 1994, **190**, 199–202.

(33) BEYAR, R., SHOFTI, R., and GRENADIER, E. Self expandable nitinol stent for cardiovascular applications: canine and human experience, *Catheterization and Cardiovascular Diagnosis*, 1994, **32**, 162–170.
(34) CRAGG, A. H., DE JONG, S. D., and BARNHART, W. H. *et al.* Nitinol intravascular stent: results of preclinical evaluation, *Radiology*, 1993, **189**, 775–778.
(35) MARNI, E. Balloon nitinol device: a new possibility for the fixation of balloons in large arteries, *Int J. Cardiology*, 1992, **34**, 224–226.
(36) ROSENTHAL, D., HERRING, M., and O'DONOVAN, T. G. *et al.* Endovascular infrainguinal *in situ* saphenous vein bypass: a multicentre preliminary report, *J. Vasc Surg*, 1992, **16(3)**, 453–458.
(37) ROSENTHAL, D., HERRING, M. B., and McCREADY, R. A. *et al.* Angioscope-assisted endovascular occlusion of venous tributaries: Preclinical results, *Cardiovasc Surg*, 1993, **1(3)**, 225–227.
(38) ROSENTHAL, D., DICKSON, C., and RODRIGUEZ, F. J. *et al.* Infraguinal endovascular *in situ* saphenous vein bypass: Ongoing results, *J. Vasc Surg*, 1994, **20**, 389–395.
(39) HUANG, J., MA, G. S., and WANG, J. L. *et al.* Experimental study on the occlusion of arterial blood flow by the implantation of nitinol alloy stents, *Chin Med J.*, 1994, **107(7)**, 512–514.

CHAPTER 5

Present and Future Orthopaedic Applications

5.1 Present orthopaedic applications of SMAs

Schettler (**1**) was the first to report the use of the shape memory effect in orthopaedic applications in 1979. Schettler used an Ni–Ti alloy to produce a constant compressive force on certain lower jaw fractures. Schettler *et al.* carried out four tests on shape memory metal plates:

(a) test with Plexiglass
(b) photoelastic test
(c) gap measurement under masticatory force
(d) osteosynthesis evaluation.

The initial two tests allowed evaluation of the attainable compressive stress gained from an Ni–Ti plate which had been screwed onto either a piece of plexiglass or photoelastic material.

The third test simulated the acts of chewing by allowing a repetitive force to be applied to the point of fracture on a flexiglass plate.

The results gained show gap size in comparison to force applied, and the eventual return after removal of load are shown diagrammatically in Fig. 5.1.

The fourth test allowed evaluation of possible Ni–Ti osteosynthesis plates. The memory plate was attached to an artificially cleved mandibular bone (Fig. 5.2).

The SMA was heated to above transition temperature and the gap was found to close (Fig. 5.3).

After cooling, the bones were opened by a stimulated masticatory force and the process of heating was repeated.

Schettler *et al.* concluded that the Ni–Ti alloy:

opens up entirely new possibilities to medicine.

The new alloy produces higher compressive stresses than the plates hitherto known and ... any subsequent opening of the fracture can be treated repeatedly and removed with the shape memory effect.

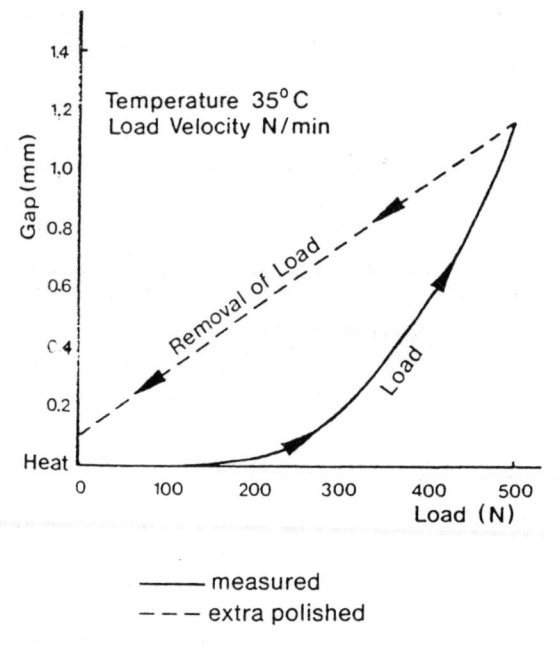

Fig 5.1 Load diagram gained by Schettler in the masticatory simulation test (1)

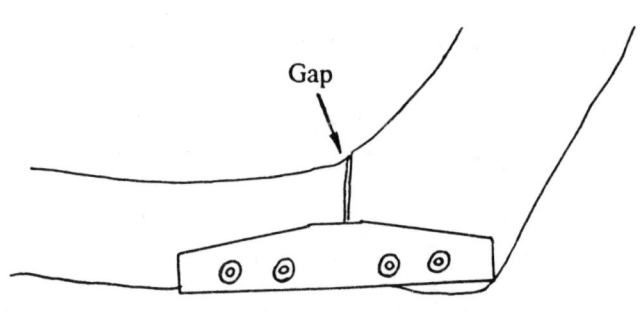

Fig 5.2 Artificially fractured jaw model used by Schettler with memory alloy plate attached (1)

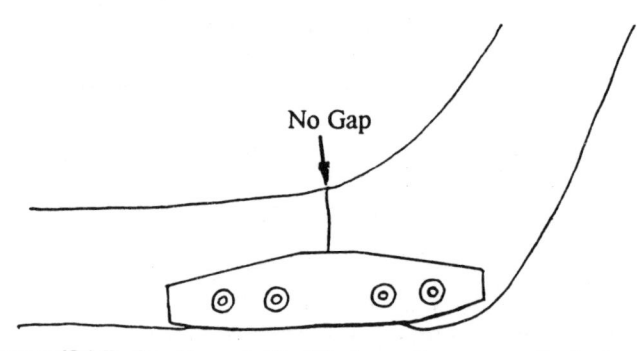

Fig 5.3 Same artificially cleved jaw as in Fig. 5.2, shown after memory alloy has been activated (1)

Since his report, further studies have been carried out into the viability of the alloy for orthopaedic operation. In 1981 a Chinese Ni–Ti shape memory alloy (TTR 34–40°C) began application in two Shanghai hospitals. Although its use was widespread throughout the hospital, in the orthopaedic department it was applied in just two forms:

(a) Ni–Ti cups: for use in surfacing arthroplasty
(b) Ni–Ti staples: used in fractures, osteotomy and arthrodesis.

Several reports have been published about investigations into these functions (2)–(6). In 1983 Yang *et al.* (4) published a paper on the initial use of Ni–Ti SMA compressive staples that were used for fracture or other orthopaedic conditions in the vicinity of joints. Yang *et al.* implanted staples into eighteen patients and early results led Yang *et al.* to conclude:

Memory alloy staples provide a new promising option of biomaterial for the orthopaedic surgeons.

In 1986, a report produced by Kuo *et al.* (3) reviewed the application of Ni–Ti shape memory alloy in orthopaedics. Kuo *et al.* concluded that:

It is believed that this new alloy is an extraordinary substance to supplement the field of biomaterials.

However, a more complete summarization of the Chinese clinical application up to 1989 was published by Kuo *et al.* (7). Kuo *et al.* reported on seventy-one orthopaedic cases, excluding dentofacial surgery, in which the alloy had been applied. The cases consisted of twenty surface arthroplasties and fifty-one applications of staples.

Surface arthroplasty involves the placement of a corrosion resistant cup on the top of the femoral head. The use of the SMA simplified this procedure by the utilization of inverted 'memorized' hooks (Fig. 5.4). At 0°C the hooks were deformed so that the cup could be positioned on the femoral head. Once set, the warming created by body temperature reverted the hooks into their original position, and in doing so, clamped the alloy cup to the bone.

After deformation at 0 degrees C

After fixation and implantation

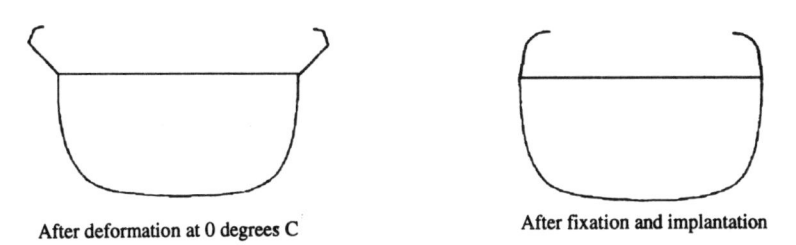

Fig 5.4 A highly diagrammatic representation of a Ni–Ti cup as used in surface arthroplasty. The hooks hold the femoral head tightly, bending the cup to the femur

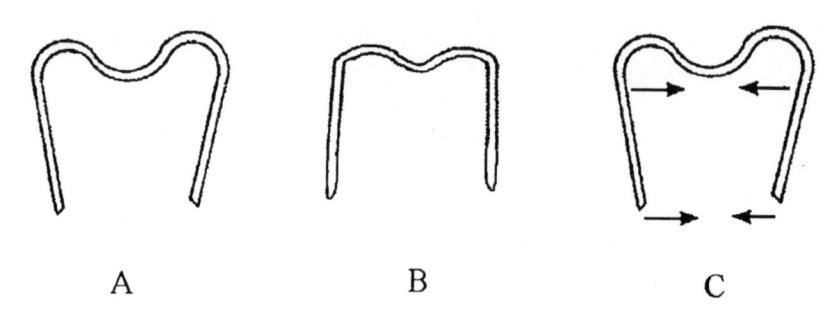

Fig 5.5 A highly diagrammatic picture of Ni–Ti orthopaedic staples as used in bone fractures (a) before deformation; (b) after deformation and at time of implant; (c) return to original form, causing a compressive force (arrows)

In cases of osteotomy, bone fracture or bone fusion, the hospitals involved used 'wavy' staples (Fig. 5.5) with 60° spikes at each end. The staples were elongated at 0°C along their U section and spikes placed within the pre-drilled holes in the bones. Upon warming by a hot saline pack (40°C), the 'wavy' section returned to its original shape and thus produced a compressive force between the two subject bones. This compressive force encouraged osterial tissue growth and hence quick fracture recovery.

Kuo et al. used a Chinese Ni–Ti alloy which could be deformed below 5°C and whose transition range was between 34°C and 40°C.

The detail of cases in which either cups or staples were utilized as described by Kuo et al. are given in Table 5.1.

Table 5.1 Orthopaedic cases treated by the shape memory alloy (Kuo et al. (3))

Treatment	Type of case	No. of occurrences
Surface arthroplasty using Ni–Ti cup	Rheumatoid arthritis	5 hips
	Aseptic necrosis	8 hips
	Old malunited fractures	5 hips
	Albers-Schönberg disease	2 hips
	Subtotal (a)	20 hips
Ni–Ti orthopaedic staples	Ankle fractures	10
	Triple arthrodesis of foot	16
	Wrist arthrodesis	4
	Hip arthrodesis	2
	Osteotomy	6
	Patella fracture	2
	Olecranon fracture	2
	Fractures of metacarpus, metatarsus & phalanx	7
	Tendon suturing	1
	Ligament suturing	1
	Subtotal (b)	51
	Complete total (a + b)	71

Kuo *et al.* commented from the results that:

We are confident that this new alloy will have a bright future.

However, the group also discussed the need for further improvement of the Chinese alloy in certain areas, such as transition temperature and obtainable compressive and extensive forces, and so on, and stated that the further investigations into the constitution of the alloy and its effect on the alloy's properties were commencing to find the 'optimal composition for clinical use'.

In 1993 Dai *et al.* (8) used similar SMA staples to treat 132 intra articular fractures. The group used staples with a transition temperature of 37°C, which were deformed at 4°C prior to implantation (Fig. 5.6). After placement, a hot saline pack was applied at the implantation site and complete recovery induced (Fig. 5.6).

Dai *et al.* found that all fractures healed naturally and no adverse tissue reaction was observed.

In the ninety-three follow up cases (average follow up = 3.1 years) highlighted by Dai *et al.* no patients' results were classed as poor and only six cases were regarded as fair, with the rest being either good or excellent (Table 5.2).

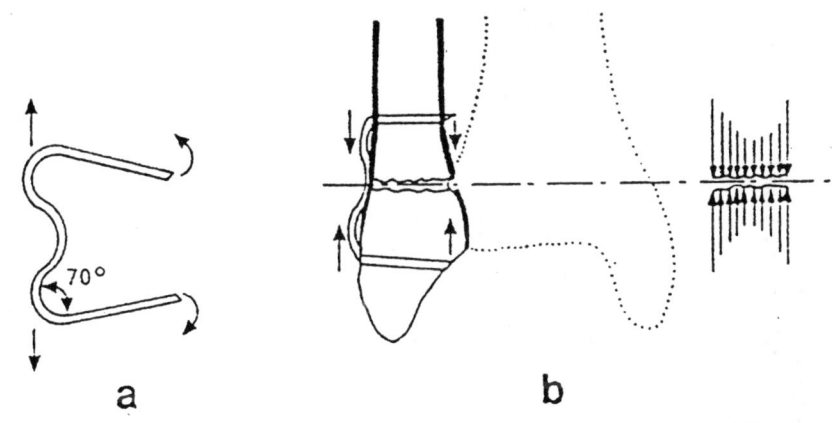

Fig 5.6 Schematic of the compressive staple used by Dai *et al.* (8). (Courtesy of Butterworth-Heinemann journals, Elsevier Science Ltd, UK.)

Table 5.2 Follow up results gained by Dai *et al.* (8)

Fracture site	Excellent	Good	Fair	Poor	Total
Patella	39	18	2	0	59
Malleolus	9	11	3	0	23
Olecranon	3	4	0	0	7
Lateral condyle of humerus	2	1	0	0	3
Tibial plateau	0	0	1	0	1
Total (%)	53 (57.0)	34 (36.5)	6 (6.5)	0 (0)	93 (100)

Dai *et al.* concluded that the NiTi SMA staple

> is exceptional in both mechanical and biological properties and has possibilities for wide medical application.

5.2 Future possible orthopaedic applications of SMAs

5.2.1 *The correction of scoliosis*

Scoliosis is by basic definition a deformity of the spine. A normal spine (Fig. 5.7) is not absolutely straight, it curves in the anteroposterior direction, but should still remain vertical.

In scoliotic backs the spine is seen to have one or more lateral curves (Fig. 5.8). These curves may vary in severity and, in some cases, cause disability or

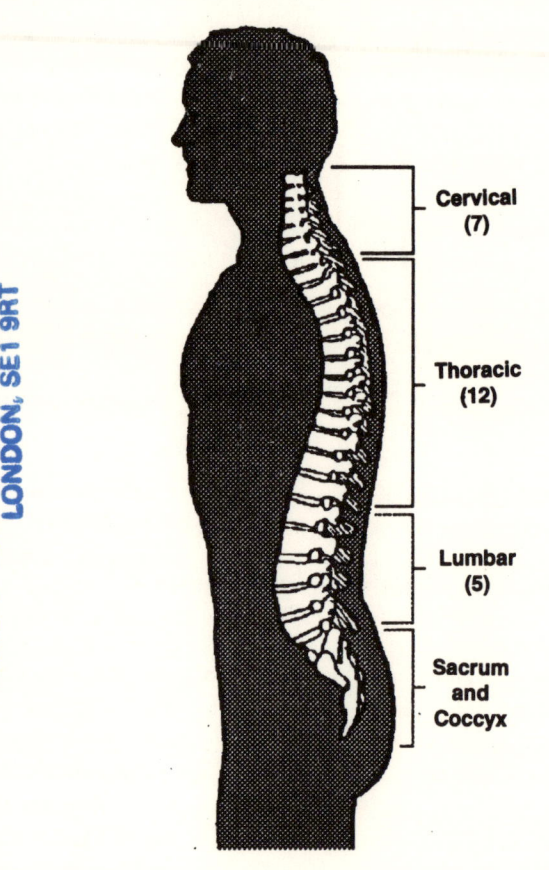

Fig 5.7 A diagrammatic representation of a normal spine (9). (Copyright (1991) IEEE.)

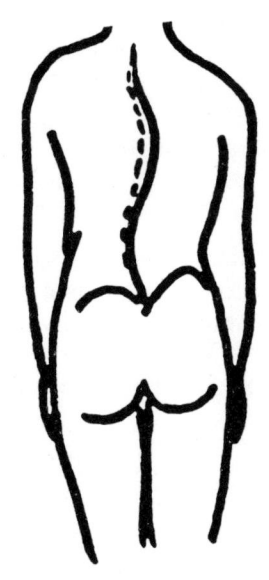

Fig 5.8 Possible lateral curvature caused by scoliosis (15). (Copyright (1988) *Journal of Biomechanics*. Courtesy of Elsevier Science Ltd, UK.)

internal organ difficulties. Treatment of scoliosis is constantly progressing with many treatments, all of which can be split into two basic groups:

(a) Non-operative: external correction of deformity commonly involving a bracing technique such as the Charleston, Boston and Milwaukee bracing systems **(10)**–**(14)**.
(b) Surgical: internal correction, usually the use of a distraction rod to correct spinal curvature.

The distraction rod technique initially involved the attachment of hooks under the laminae of vertebrae above and below the curve to be corrected. The spine is then longitudinally straightened by an external device and a distraction (Harrington) rod attached to the hooks. Bone material is then applied around the spine to provide a basis for osteosial growth and eventual bone fusion in the straightened position. However, measurements have shown that the correcting force falls to approximately 30 percent of its original value after 10–15 days and to re-establish the initial force after this period of time requires a second operation to be performed.

A group led to Schmerling **(16)** carried out a feasibility study on a new type of distraction rod, made from shape memory alloy, that could overcome the need for surgical correction. Schmerling *et al.* proposed the use of a rod made out of Ni–Ti alloy ($A_f = 43°C$). Implantation of the device would occur in the usual fashion with the rod's memory shape being straight. On placement, the same reduction of forces as previously described, would occur. However, after a set period of time the alloy may be externally heated above A_f and hence transform into its preset shape and exert the same, or nearly the same, force as originally applied. Schmerling *et al.* used an *in vitro* system for testing and, although the group found difficulties with heating the alloy in an even manner, they concluded that:

> The shape memory effect in near equiatomic nickel–titanium has been used to advantage in the Harrington rod treatment of scoliosis. An Ni–Ti Harrington rod, originally straight but with an imparted curvature, has the ability of straightening to its original shape and post-operatively re-establishing the relaxed corrective forces. This can be accomplished by external heat application.

However, it wasn't until nearly twenty years later that any further interest was shown in the use of SMAs to correct scoliosis. In 1993 Matsumoto *et al.* **(17)** produced a paper describing their investigation into the gradual correction of artificially induced scoliosis in two monkeys using L-shaped (dia = 5 mm) NiTi SMA rods (Fig. 5.9).

The group implemented radio frequency induction heating to induce complete recovery (TTR = 35°–48°C), and Matsumoto *et al.* concluded that:

> Adequate restoration power for correction of scoliosis was obtained using alloy rods with a diameter of 5 mm.
>
> The spine could be gradually corrected using this instrument [NiTi rods].

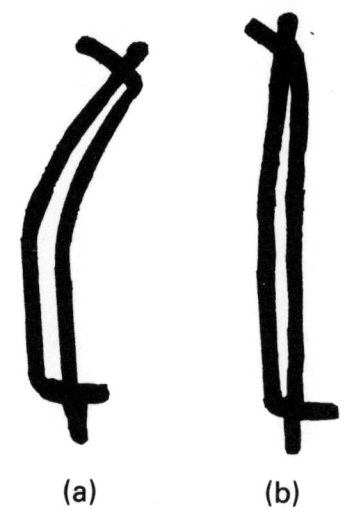

Fig 5.9 L-shaped rod used by Matsumoto for stepwise correction of scoliosis (a) before recovery, (b) after recovery (17)

The correction rate of the spine could be accurately controlled by control of the temperature increase on the alloy.

A further report published in 1993 by Sanders *et al.* (**18**) discussed the implantation and correction of experimental kyphtotic scoliosis in six goats using 6 mm diameter NiTi SMA rods.

The rods were deformed in iced saline and implanted. Shape recovery was induced using an external radio frequency heater, and any symptomatic complications during rod recovery were noted.

Sanders *et al.* observed that none of the goats showed signs of discomfort during the heating process, and all appeared to have suffered no form of neurological or thermal injury to the spinal cord or surrounding tissue.

In the histological evaluation after five months Sanders *et al.* found there to be

> no evidence of severe injury to the tissues or spinal cord in the goat.

They also found

> there was no evidence of a significant foreign body response.

Sanders *et al.* also commented that the nitinol alloy offered

> the possibility of correcting spinal deformity by taking advantage of spinal viscoelasticity. A high margin of neurologic safety is affected as well because the correction is performed in an awake patient.

The group concluded that with qualities discussed this

> may make shape memory alloys very useful in scoliosis.

5.2.2 The union of fractures

In 1989 a further investigation into fracture fixture was published by Zhang (**19**). Zhang published a report which involved using a shape memory fixture to restore communicated patella to their original shape. Zhang implanted the Ni–Ti fixture into thirty-three cases and found that, after only approximately seven weeks, the patient had gained full extension and flexion of the knee joint. The Zhang study group all regained good functional ability of the leg without complication.

In 1992 Yang *et al.* (**20**) reported on a novel clamp for internal fixture of fractured tubular bones. The clamp consisted of one main rod with three or four pairs of curved arms protruding from both sides (Fig. 5.10), depending on what was suitable for the fracture. The arms were opened in sterile ice water and upon heating above its transition temperature the arms would rotate into a circular position, and hence clamp the bone around approximately $\frac{2}{3}$ of its circumference. Over a three year period, Yang *et al.* implanted the clamps into sixty-four patients with a follow up study group of fifty-two people (male $n =$

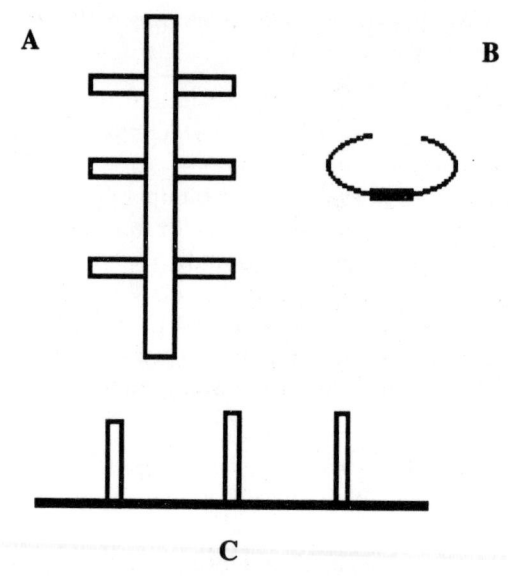

Fig 5.10 A highly diagrammatic representation of the shape memory clamp implant used by Yang et al. (20). (a) top view; (b) front view; (c) side view

33, female $n = 19$), whose mean age was twenty-nine years. The fractures treated within the study group are given below in Table 5.3.

A cast producing external immobilization was utilized for an average of 3.5 weeks after implantation, and no malunion or non-union occurred. Histological investigation was carried out on the non clamped and clamped regions.

Yang et al. concluded of his case study:

> The fracture healed excellently without bone resorption. Pathological examination showed that the specimens taken from both compressed and uncompressed sites of the fragments showed no histological difference.

The clamping performed by Yang allowed fixture of the bone at many points, which is advantageous in the quick recovery of tubular breaks. The new device was especially encouraging as it showed good potential for fixing communited

Table 5.3 The classification of the fracture (8)

Types	Cases	Locations	Cases	Lines	Cases
Fresh closed	27	Hand phalanx	14	Transverse fracture	5
Fresh open	20	Hand metacarpal	18	Oblique fracture	21
Old	3	Foot digital	2	Spiral fracture	3
Nonunion	2	Foot metatarsal	3	Comminuted fracture	23
		Other	15		

fractures of short tubular bones, a treatment that is difficult to perform using other methods.

5.2.3 The use of orthotics

In 1992, Takami *et al.* (**21**) produced an initial report into the use of the nickel–titanium shape memory alloy for hand splinting. Takami *et al.* applied the alloy for two types of splint:

(a) reverse knuckle bender splint
(b) cock-up splint.

The group found the shape memory splints easy to attach and cooled below their TTR. The gradual recovery due to room or body heat avoided the development of spasticity. Takami *et al.* concluded that:

> The splint was easy to wear and could be worn with comfort for an extended period.
> The authors especially emphasize the point that using a shape memory alloy as the assist or extension bar of a splint may be the most desirable method of correcting a deformed hand with spasticity.

In quadriplegic patients (Table 5.4) the ability to perform nominal activities of daily life (ADL) are often seriously constrained and, although various assistive devices are available to enhance the subject's independence, such devices are often unsightly and heavy due to the type and nature of the materials used.

Table 5.4 Functional independence of subjects with spinal cord trauma

Level of damaged segment	Functional independence after damage
C^{4}*	Type, turn pages, use of the telephone and computer with a mouth stick.
C^{5}	All C^4s activities plus
	Feed
C^{6}	All C^5s activities plus
	Drink
	Wash, shave, brush hair
	Dress upper half
	Sit up/Lie in bed
	Write
C^{7}	All C^6s activities plus
	Turn in bed
	Dress lower half
	Skin care
C^{8}	All C^7s activities plus
	Bladder and bowel control

*C^4–C^8 distinguish the position of spinal damage, C indicates trauma is in the cervical vertebrae region, and 4–8 represent which vertebra in question. (see also Fig. 5.7)

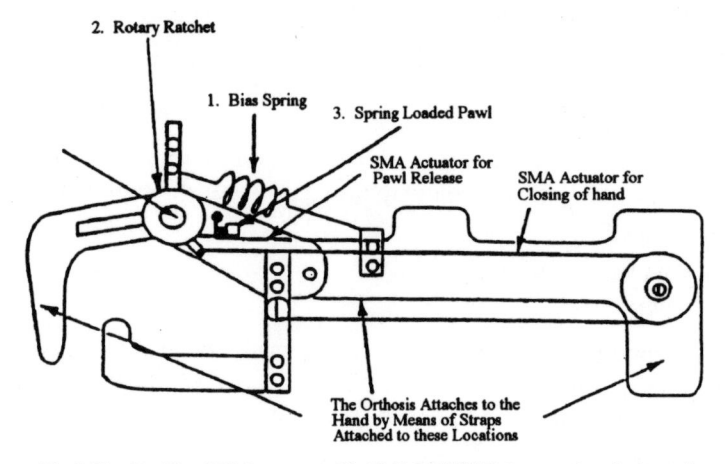

Fig 5.11 Profile of Makaran et al.'s SMART WHO for quadraplegic patients

In 1993 Makaran et al. (22) discussed their prototype 'SMART®' wrist–hand orthosis (WHO) which utilized the shape memory properties of NiTi as a lightweight actuating element in the opening and closing of a subject's hand (Fig. 5.11).

On the patient initiation the 'SMART' orthosis implements electrical resistant heating, from a small battery pack, to tighten the SMA Actuator, the rotary rachet (labelled 2. in Fig. 5.11) assuring the hand position is maintained. Once object release is required the second NiTi actuator is activated pulling the spring loaded pawl (labelled 3. in Fig. 5.11) away from the rotary ratchet and allowing the bias spring to return the hand to its original position. The orthosis was tested on a C5 quadraplegic who had use of his deltoid and bicep muscles. The group found the device allowed the patient the ability to use ordinary everyday objects, and the subject verbalized his endorsement of the orthosis.

Makaran et al. concluded that although improvements to the prototype were possible the 'SMART' orthosis design was

> lightweight and simple, combining ease of use and good functionality.

They went on to conclude that

> The SMART WHO prototype provides a new concept in the design of wrist–hand orthosis, allowing for greater independence in ADL.

5.3 Conclusion

The investigations into SMA orthopaedic applications were initiated with a preliminary report evaluating the potential of the Ni–Ti alloy on simulated fractures of the lower jaw in 1979. Due to its relative recent research basis, the

SMA has fewer reports and even fewer applications to which it has been applied: orthopaedic staples, cups and clamps have been developed, possible Harrington rods for scoliosis have been visualized, and hand/wrist splints have been investigated.

Chinese research leads the world in studies of this kind, with clinical results already gained from the application of compressive staples and surface arthroplasty cups, from two Shanghai hospitals. Other study areas such as splints, clamps, and so on, are within their investigational infancy, though initial reports have been very encouraging.

The Ni–Ti SMA shows advantages over other more conventional solutions. In surface arthroplasty, it enables relatively easy placement onto the femoral head, and its stability once in position has been seen to be good (7). Compressive staples enable repair of fractures via an *in vivo* method, cutting down the possibility of entry sight infection and lowering the discomfort felt by patients. The staples show good mechanical qualities and bones have been seen to repair quickly and efficiently (7).

The other orthopaedic applications have not advanced to the stage of clinical trials at present, though all show great promise within their chosen fields.

The large advantages possible, and the ease of application obtainable, make the SMA a potentially valuable weapon within the orthopaedic surgeons' armoury, and a possible solution to various non-operatic orthopaedic conditions.

References

(1) SCHETTLER, D. Method of Alveolar bracing in mandibular fractures using a new form of fixation made from memory alloy, *J. Maxillofacial Surg*, 1979, **7**, 51–54.

(2) ZHANG, Y. F., TAO, J. C., and CAI, T. D. *et al*. A preliminary report on application of Ni–Ti 'Memory' alloy for triple orthodesis of foot, *Acto Shanghai 2nd Univ of Med Sciences*, 1985, **2**, 149.

(3) KUO, P. P. F. *Clinical use of nickel–titanium shape memory alloy in orthopaedics surgery, in progress in artificial organs*, 1986 (ISAO Press), 1105–1107.

(4) YANG, P. J., ZHANG, T. F., and KUH, M. Z. *et al*. The use of Ni–Ti shape memory alloy staples for internal fixation of fractures, *Chinese J. Orthop*, 1983, **3**, 137–140.

(5) DAI, K. R. The use of compressive staples of Ni–Ti shape memory alloy in orthopaedic surgery, *Chinese J. Orthop*, 1983, **21**, 343–348.

(6) YANG, P. J., ZHANG, Y. F., and YAO, J. C. et al. Internal fixation with Ni–Ti shape-memory alloy compressive staples in orthopaedic surgery: a review of 51 cases, *Chinese Med J.*, 1987, **100(9)**, 712–714.
(7) KUO, P. P. F., YANG, P. J., and ZHANG, Y. F. et al. The use of nickel–titanium alloy in orthopaedic surgery in China, *J. Orthop*, 1989, **12(1)**, 111–116.
(8) DAI, K. R., HOU, X. K., and SUN, Y. H. et al. Treatment of intra-articular fractures with shape memory compressive staples, *Injury*, 1993, **23(10)**, 651–655.
(9) GOEL, V. K. Effects of Injury on the Biokinetics of the lumbar spine, *IEEE Eng in Med and Bio*, June 1991, 42–47.
(10) EVANS, J. B., KAELIN, A., and BANCEL, P. et al. The Boston bracing system for loliopathic scolosis: follow up results in 295 patients, *Spine*, 1986, **11(8)**, 792–801.
(11) CHASE, A. P., BADER, D. L., and HOUGHTON, G. R. The biomechanical effectiveness of the Boston brace in the management of adolescent loliopathic scolosis, *Spine*, 1989, **14(6)**, 636–642.
(12) PRICE, C. T., SCOTT, D. S., REID, F. E., and RIDDICK, M. F. Nighttime bracing for adolescent loliopathic scolosis with the Charleston bending brace: Preliminary report, *Spine*, 1990, **15(12)**, 1294–1299.
(13) KOSTUIK, J. P., CARL, A., and FERRON, S. Anterior Zielke Instrumentation for spinal deformity in adults, *J. of Bone & Joint Surg*, 1989, **71-A(6)**, 89–91.
(14) POPE, M. H., STOKES, I. A. F., and MORELAND, M. The biomechanics of scoliosis, *CRC Crit Rev Biomed Eng*, 1985.
(15) GHISTA, P. N., VIVANI, G. R., and SUBBARAJ, K. et al. Biomechanical basis of optimal scoliosis surgical correction, *J. Biomechanics*, 1988, **21(2)**, 77–88.
(16) SCHMERLING, M. A., WILKOR, M. A., SANDERS, A. E., and WOOSLEY, J. E. A proposed medical application of the shape memory effect: An Ni–Ti Harrington rod for treatment of scoliosis, *J. Biomed Mater Res*, 1976, **10**, 879–902.
(17) MATSUMOTO, K., TAJIMA, N., and KUWAHARA, S. Correction of scoliosis with shape memory alloy, *J. Jpn Orthop Assoc*, 1993, **67**, 267–274.
(18) SANDERS, J. O., SANDERS, A. E., and MORE, R. et al. A preliminary investigation of shape memory alloys in the surgical correction of scoliosis, *Spine*, 1993, **18(12)**, 1640–1646.
(19) ZHANG, C. C. Treatment of patella fracture using an internal fixation of shape-memory alloy, *Chung-Hua Wai Ko Tsa Chih (Chinese)*, 1989, **27(11)**, 692–695.
(20) YANG, P. J., TAO, J. C., and GE, M. Z. et al. Ni–Ti memory alloy clamp plate for fracture of short tubular bone, *Chinese Medical Journal*, 1992, **105(4)**, 312–315.

(21) TAKAMI, M., FUKUI, F., and SAITOU, S. *et al.* Application of a shape memory alloy to hand splinting, *Prosthetics & Orthotics Int*, 1992, **16(1)**, 57–63.

(22) MAKARAN, J. E., DITTMER, D. K., and BUCHAL, R. O. *et al.* The SMART Wrist-hand Orthosis (WHO) for quadriplegic patients, *J. Prosthetics Orthotics*, 1993, **5(3)**, 28–30.

CHAPTER 6

Other Applications of the SMA

6.1 Dental arch wires

6.1.1 A brief history of arch wires

Orthodontic treatment is based on the systematic application of forces. Forces are produced in arch wires due to the elastic properties of metals. Once deformed, these properties are displayed as a natural tendency to return to the wires' original shape, and the magnitude of the returning force – the elasticity – is dependent on the type and size of material used. In orthodontics these returning forces are applied to 'tug' misaligned teeth into their optimal positions.

In the early days of orthodontics gold alloy was the most commonly used material for dental arch wires (**1**)–(**4**). However, due to its increasing cost, its lack of elasticity and the appearance of stainless steel in the 1940s, the use of gold alloy declined.

Arch wires utilized the materials' elastic capabilities and to maximize this, the wires were made thinner and changed from a rectangular to a round type. The development of loop techniques (**5**)–(**12**) since the mid 1950s allowed a further reduction in the stiffness of the wire and an increase in its working range. Braiding or twisting stainless steel allowed still further increases in the arch wires' working range. In 1976, a new orthodontic wire made from nitinol (a Ni–Ti alloy) was introduced. Since its introduction a number of commercial applications use the hard worked nitinol alloy, which is unaffected by temperature, but portrays superelasticity. More recently, temperature dependent shape memory, nitinol and other Ni–Ti alloys have been developed.

6.1.2 SMA thermal arch wires

Since their first feasibility study in 1971 by Andreasen and Hilleman, wires made of the nickel titanium alloy have gained much popularity in the orthodontic field. Although initial investigations involved the superelasticity property of

Table 6.1 Dimensions and transition temperature range of wire used by Andreasen and Brady (13)

Dimensions	Nitinol wires used by Andreasen & Brady		Transition temperature range of wire °C
	Original length	Stretched length	
0.02 round	100-mm	108-mm	16–27
0.02 round	100-mm	108-mm	32–42

the Ni–Ti alloy, other investigations have been carried out into its thermomechanical property.

Andreasen and Brady (13) first discussed a use of the shape memory phenomenon in 1972, a year after Andreasen's *et al.* initial evaluation. Andreasen and Brady stretched two differently characterized (that is with differing TTRs) Ni–Ti wires (Table 6.1) and recorded the shrinkage force created as the temperature was raised. The results are shown diagrammatically in Fig. 6.1.

From their results, Andreasen and Brady concluded that:

> It appears a range of forces can be selected... The range of forces for a 108 mm stretched wire would be from approximately 1.5–5.0 lb between the limits of room and body temperature.

Andreasen and Brady were the first to show that by altering the materials' TTR a variable force could be applied.

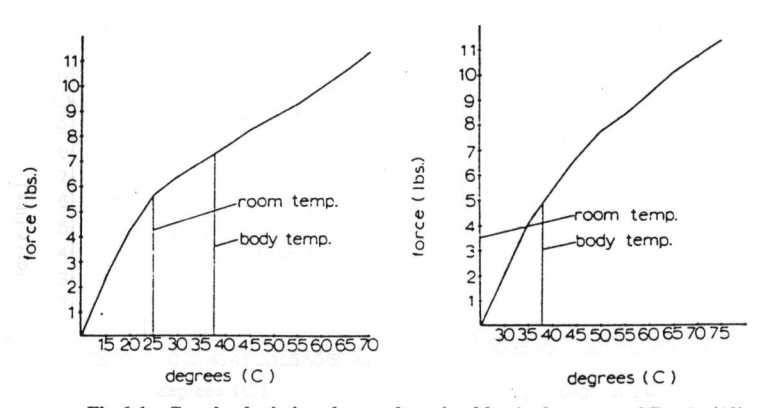

Fig 6.1 Graphs depicting the results gained by Andreasen and Brady (13)

In 1979 Andreasen, Bigelow, and Andrews (**14**) suggested using an Ni–Ti arch wire to close extraction spaces. When treating a large number of malocclusions, there may be a need to remove two bicuspids or other teeth in each dental arch. The remaining teeth must then be moved to close the extraction space or spaces. A common method of achieving this is by utilizing cemented edgewire brackets and attaching a parabolic shaped arch wire. This allows the teeth to be brought into positions of interproximal contact. Andreasen *et al.* tested three wires with differing TTRs at three different temperatures (Table 6.2).

Andreasen, Bigelow, and Andrews investigated the force produced as a function of recovered length. The wire was stretched until a resultant plastic deformation of 8 percent occurred. The temperature was then raised to a preset value (shown in Table 6.2) and the recovery force, and recovered length, recorded. Andreasen *et al.* commented in their report that:

The alloy has several qualities which makes it excellent for orthodontic space closure application.

They also concluded that the Ni–Ti alloy could close up the majority of two tooth extractions with only one change of wire. There are five or six changes of wire required for such a closure using stainless steel, and so it can be seen that using Ni–Ti alloy is very advantageous.

The group added:

At most it would take two changes of pre-stretched wire to close an extraction site.

In 1980 Andreasen (**15**) reported on a clinical trial using 0.019 inch Ni–Ti wire with a transition temperature between 31°C and 45°C to realign teeth in the mandibular dental arch.

Within orthodontics the antero-posterior relationship between the upper and lower dental arches, ignoring all lateral deviation or malpositioning, is described by a set of classes based on those defined by Edward Angle in 1899 (**16**).

Table 6.2 Test wire used and temperature tested at (Andreasen *et al.*, 1979) (14)

Number of wires	Temperature tested at (°C)	Transition temperature range of wire (°C)
2	39	15.5–27
2	39	37–40.5
2	39	43–49
2	45	15.5–27
2	45	37–40.5
2	45	43–49
2	50	15.5–27
2	50	32–40.5
2	50	43–49

Note: All wires were at the manufacturer's specifications

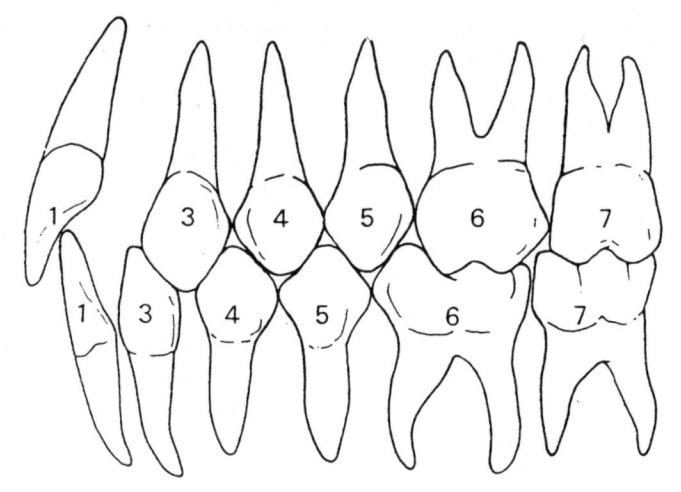

Fig 6.2 Diagrammatic examples of Class 1 occlusion. The accepted ideal occlusal relationship (17)

A relationship (Fig. 6.2) (**17**) is classified as a Class 1 when the teeth of both jaws are aligned within their correct 'ideal' places. A Class 2 relationship at a very basic level can be described as being a situation where the lower dental arch is in a more posterior position in relation to the upper arch, and a Class 3 relationship is where the lower arch is in a more anterior position relative to the upper dental arch.

In Andreasen's report, the patient – a thirty-two-year-old female – was diagnosed having a left central incisor that was lingually blocked out, the lower lateral incisors were crowded and partly lingual to canines. However, the patient had a Class 1 intercuspation between canine and molar.

The treatment required the left central incisor to be extracted, followed by the realignment of the three remaining mandibular incisors.

The gradual progression from lingual positions to final alignment was shown photographically by Andreasen *et al*.

Andreasen found that the use of the Ni–Ti arch wire realigned the teeth in 163 days and demonstrated the effectiveness of Ni–Ti wire for at least initial teeth alignment.

Andreasen concluded that:

> The clinical trial described in this article demonstrated that 0.019 inch thermal nitinol wire, when activated by the heat in the mouth, moved the mandibular incisor into normal crown alignment.

Other reports since 1980 (**18**)–(**27**) have, on the whole, concentrated on the superelasticity of the annealed Ni–Ti alloy and advancements in this area. However, in 1989 Andreasen and Amborn (**28**) reported on a pilot study of four subjects, three of whom needed no extractions and the fourth the removal of four premolars. The purpose of the study was to show the capacity of the Ni–Ti alloy wire to serve as a levelling and aligning arch wire that had torque control. Each patient was fitted with thermal arch wire in the maxillary dental arch and stainless steel (multi strand) in the mandibular dental arch. Each patient was seen weekly for a period of sixteen weeks and any complaints of discomfort were recorded. Andreasen and Amborn concluded that:

> These clinical trials demonstrated that the 0.017 × 0.025 thermal Ni–Ti wire is effective as an initial alignment and levelling arch wire with torque control over the roots, even in cases with severely malpositioned teeth ... No differences in discomfort between thermal Ni–Ti wire, twisted wire and stainless steel wire was noted by patients.

The development of thermal 'shape memory' Ni–Ti wire as an arch wire has progressed. Features affecting this progression have been the costs of production and the difficulties faced in maintaining consistency of the TTR in the manufacturing process.

It has, however, been shown by clinical trials that the alloy is an effective arch wire for aligning and levelling dental defects. Some more recent reports (**24**) (**26**) have discussed the feasibility of recycling the Ni–Ti arch wires and the effect it has on the mechanical properties of the alloy. These reports have been based on the superelasticity form of the alloy. If a recycling programme could be involved with the thermal Ni–Ti wire this may enable a lowering of costs and hence the wider application by practitioners.

6.2 Other thermally activated SMA stents

6.2.1 Introduction

The use of nickel–titanium activated stents in the intravascular system, following percutaneous transluminal angioplasty has been previously described. However, vascular endoprosthesis are not the only form of stenting. Other types have been investigated in recent years, including biliary (**29**), tracheobronchial, esophageal (**30**), ureteral (**31**), and rectal, although until recently only tracheobronchial (**32**)–(**34**) had shown any major connection with shape memory alloys.

6.2.2 Tracheobronchial stents

The occlusion of a major airway may be due to several factors such as intraluminal disease, stenosis, and so on. In some cases surgical reconstruction is possible (**35**). However, an alternative in some cases can be found by using intraluminal stents.

Rauber *et al.* (**36**) reported on the feasibility of thermally activated tracheobronchial stents in 1990. The group perorally introduced twelve prostheses into the trachea of twelve normal rabbits. From their results Rauber *et al.* concluded that:

> Perorally insertable prostheses made of the shape memory alloy Ni–Ti may be used as endotracheal or endobronchial prostheses.

In 1991 Nakamura *et al.* (**37**) implanted ten stents into ten dogs. The stent consisted of a SMA wire (TTR – 20°C), diameter 0.9 mm made into a 30 mm long prostheses with a horseshoe cross-section allowing the airway to be sustained at the cartilage only. A small section of tracheal cartilage was removed from the animals to simulate a tracheomalasia model in accordance with Hanawa (**38**). The stents were straightened under liquid nitrogen and implanted perorally into the study region. The animals were sacrificed and examined at one week and six months. Nakamura *et al.* found that nine out of the ten stents were located in the correct region and these showed an epithelium covering of, on average, 50 percent of their surface after six months.

In 1992 Nakamura *et al.* (**39**) reported on a similar experiment. This time they used a 40 mm stent, whose transition temperature was 25–30°C and consisting of 0.5 mm diameter wire with a prosthetic outer diameter of 15 mm. The group used a thermal stent which was encased by a 300 μm silicon coating.

The wire's memorized shape was that of a zig–zag or horseshoe arrangement for the same reasons as stated above.

The endoprosthesis was deformed in ice water and perorally implanted into ten 8–14 kg adult mongrel dogs who had had a section of cartilage removed to form the tracheomalasia model.

Follow up autopsies were made at two weeks, 1, 2, 3, 4, 5, and 6 months, with two dogs left alive for a long-term study.

Nakamura *et al.* concluded that:

> ... Further structural improvements seem to be required to avoid stress concentration on airway tissue.

However, the group further concluded that despite this drawback:

> This transluminal technique for preserving airway patency shows promise for clinical application.

In 1994 Vinograd *et al.* (**40**) discussed the possible pediatric application of a nitinol removable tracheobronchial stent. The group perorally inserted twenty nitinol stents, constructed from a flat Ni–Ti sheet about 1.8 mm thick and 0.18 mm wide into twenty young rabbits. At room temperature the sheet (transition temperature <40°C) was transformed into a spiral shape (diameter 7.2 mm, length 3.3 mm) and electrical wire connections made to its two edges.

The temperature was then lowered to between 4–6°C and the stent compacted to an external diameter of 2.2 mm (length 4.2 mm). After positioning at the placement site with the aid of a bronchioscope, the stent was warmed to 40°C using electrical means and the bronchioscope removed (Fig. 6.3).

Other Applications of the SMA 113

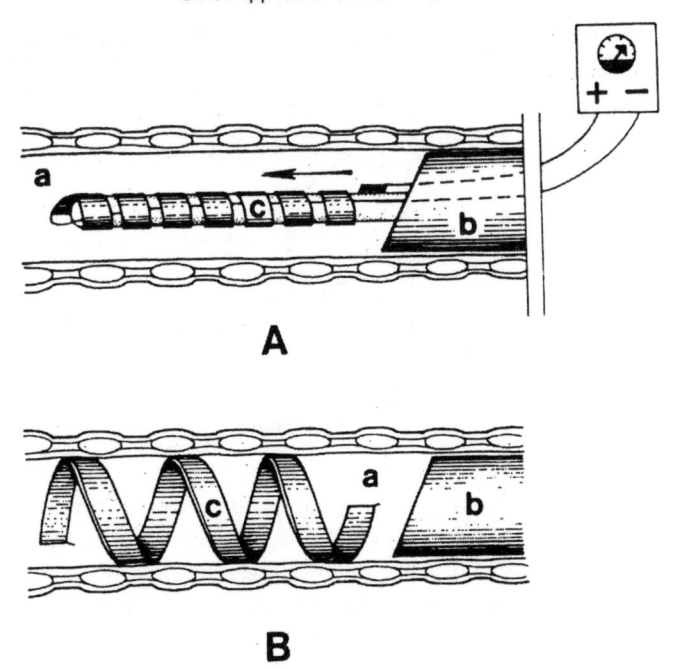

Fig 6.3 Schematic illustration showing the technique of implant used by Vinograd *et al.* (40)

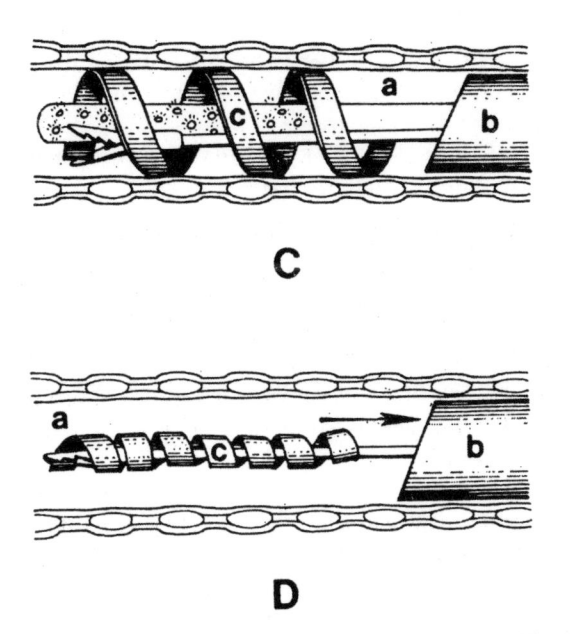

Fig 6.4 Schematic illustration showing the technique of removal used by Vinograd *et al.* (40)

Two animals were lost to follow up and the remaining animals were sacrificed after 8–10 weeks. In ten rabbits Vinograd *et al.* removed the stent with the aid of a grasper and 2 to 3 ml of cooled 80 percent alcohol solution (Fig. 6.4) prior to sacrifice with the rest being killed with the stent *in situ*.

The group found that all the stents were 'well tolerated in the animals' and concluded that:

> by taking advantage of its unique feature of shape memory effect the stent can be inserted, fixed and removed easily, even in a very small airway.
>
> In infants and children with very narrow airways, the nitinol intratracheal stent may be used as a supplementary method of achieving temporary airway splinting.

6.2.3 Urethral stents

The concept of urethral stenting arose as an offshoot from cardiovascular endoprosthesis investigations, initiated by Dotter in 1969 (**41**). However, it wasn't until 1988 that the first urethral investigation was published by Milroy *et al.* (**42**). Since that time many papers have been written on the subject, using various materials, but the relatively recent advent of shape memory alloys has meant that their application in this field has been slow.

Urethral stenting is involved in the management of two main problem areas:

(a) urethral strictures
(b) benign prostatic hyperplasia.

In 1993 Yachia (**43**) reviewed the Hillel Yaffe medical centre's experience with nitinol stents in relation to other available intra-urethral metallic stents in the management of urethral strictures.

Urethral strictures are caused by the proliferation of scar tissue along the area of the urethra, commonly after inflammatory injury.

Conventional management of such strictures involves either progressive dilation, similar to balloon angioplasty in the vascular system, or optical urethrotomy followed by catheteral or plastic material temporary stenting for a period of up to three weeks. However, these conventional treatments are often noncurative and sometimes require frequent repetition.

Yachia investigated three available types of metallic intra-urethral prosthesis:

(a) urolome/wallstent
(b) titanium ASI stent
(c) nickel titanium (urocoil) stent.

The characteristics and placement areas of each type is described in Table 6.3.

The restrictions that constrain both the urolome/wallstent and ASI stent do

Table 6.3 Characteristics and placement sites for the three types of stent highlighted by Yachia (43)

Type	Flexibility	Expansion	Stent design	Placement type: permanent/temporary	Areas of application
Urolome/Wallstent	Good	Self	Mesh	Permanent	Non mobile parts of urethra – posterior or bulbar
ASI	Poor	Balloon	Mesh	Permanent	Posterior urethral strictures only
Urocoil	Good	Self (SMA)	Eight coils	Temporary	Entire length of urethra

Table 6.4 Description of placement areas for the three SMA stents used by Yachia (43)

Design	Stricture, design aimed at
Basic urocoil	Any strictures situated between the mid-bulbar urethra and the urethral meatus
Urocoil-S	Intended for bulbo-membranous strictures
Urocoil-Twin	Intended for combined strictures of the prostatic and bulbo-membraneous urethra

not affect the urocoil due to its high flexibility and smoothness. However, due to the anatomical changes along the urethra, three designs are needed to satisfy the differing requirements (Table 6.4, Figs 6.5, 6.6, 6.7).

Yachia implanted sixty-five stents into fifty-six patients; nineteen urocoil, twenty-seven urocoil-S, ten urocoil-twin. Although he gave no details of insertion or removal, Yachia commented that:

> No patient developed clinical infection caused by the stent and none of them became occluded by stone development in or on the stent.

Yachia discussed the fact that the SMA nitinol stent could overcome three contraindications of the other types of stents:

(a) strictures after urethroplasty
(b) stricture developing after anastomotic urethroplasty performed after traumatic rupture of the posterior or the bulbo-membraneous urethra
(c) penile urethral strictures.

Yachia concluded that:

> The results obtained using the intraurethral stents are very promising. Probably the insertion of a stent will replace frequent urethral dialations on urethrotomies and ease the misery of the patients.

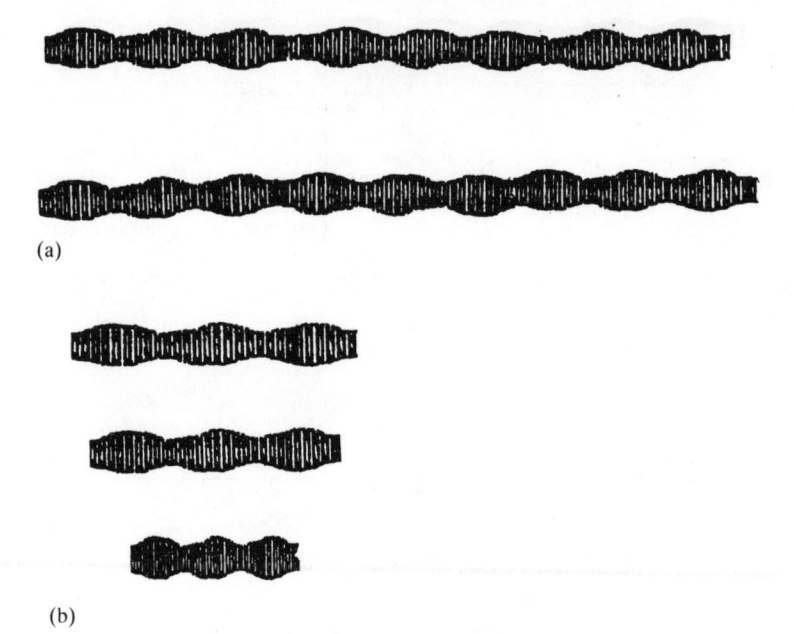

Fig 6.5   The urocoil stent as used by Yachia (43)

Fig 6.6   The urocoil-s stent as used by Yachia (43)

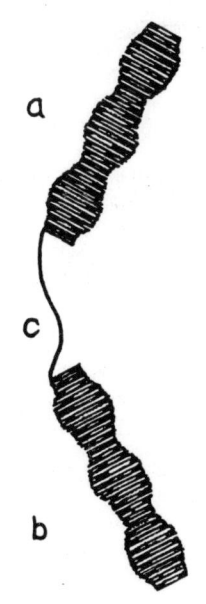

Fig 6.7 The urocoil-twin stent as used by Yachia (43)

In another report published in 1993, Poulsen *et al.* (**44**) discussed the implementation of a prototype NiTi ('Memokath') SMA spiral, to manage patients with prostate outflow obstructions.

Poulsen *et al.* implanted the 'Memokath' prostheses (Fig. 6.8, Fig. 6.9) using a modified hot saline technique (see cardiovascular section – Chapter 4, p. 72) into thirty patients with ureteral obstruction due to benign prostatic hyperplasia. The results gained are shown in Table 6.5.

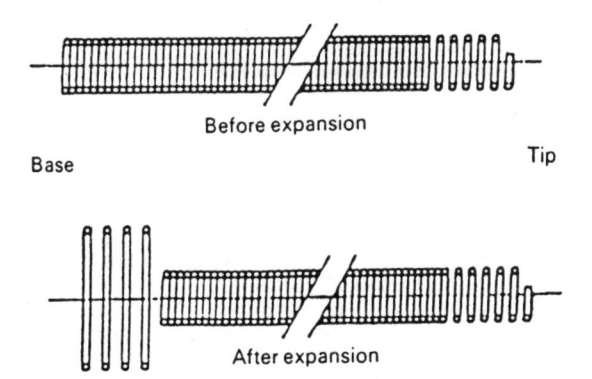

Fig 6.8 Schematic drawing of Poulsen's *et al.* spiral, before and after expansion (44)

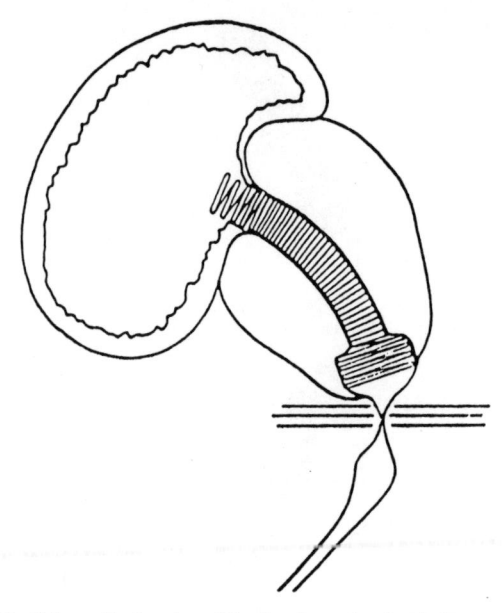

Fig 6.9 Schematic drawing of Poulsen's *et al.* spiral in its correct position (44)

No problems of migration were encountered and although a few cases showed minor complications, the group concluded that:

Treatment with the memokath seems promising.

Male spinal-cord-injured patients pose problems otherwise unseen in normal urethral stenting; there may be trauma in the movement to and from the bed/chair which might cause a migration of a stent.

In 1994 Soni *et al.* (**45**) reported on the implantation of the aforementioned NiTi memokath endoprosthesis into ten spinal cord injured males suffering with urine retention caused by detrusor-sphincter *dyssynergia*.

Table 6.5 Voiding results of the *in situ* SMA spiral as observed by Poulsen *et al.* (44)

Symptoms	None No. of patients	Slight No. of patients	Moderate No. of patients	Severe No. of patients
Frequency	15	9	1	5
Urgency	21	6	1	2
Stress and/or urge incontinence	29	0	0	1
Nocturia	7	14	9	0
Obstructed voiding	26	0	1	3
Perineal discomfort	30	0	0	0

The table includes patients whose spirals were removed because of side effects or residual symptoms.

The group's 'memokath' prosthesis was a spiral which, when flushed with hot water (above 45°C), reverted to its memorized shape. The design also incorporated the ability to make the prosthesis 'super soft', and thus removable, by lowering the ambient temperature to less than its martensite transformation temperature. This was achieved by the flushing of the placement site with cool (<10°C) water. The profile of all the patients listed is shown in Table 6.6.

Soni *et al.* found that

No patient developed profuse urethral bleeding or haematuria.

The group discussed its possible role as a temporary stent due to its nonincorporation into the mucosal layer of the urethra, and commented that the memokath stent could be used for preserving the fertility of the patient.

The group concluded that in general

Urethral stenting represents a major therapeutic innovation in the management of the neuropathic bladder.

In 1994 Qui *et al.* (**46**) investigated the application of NiTi SMA urethral stents in elderly patients with benign prostatic hyperplasia and associated diseases. The subjects were all contraindicated from surgery due to their age and health and would have conventionally undergone long-term indwelling bladder catheterization with its high risk of infection and patient discomfort.

Qui *et al.* implanted conical stents into twenty-five patients aged between 55 and 82 years with the length of stent ranging between 50 ($n = 8$), 55 ($n = 10$), and 65 mm ($n = 7$) each with a large and small diameter of 8 mm and 7 mm respectively (Fig. 6.10). The prosthesis memorized shape was activated once at

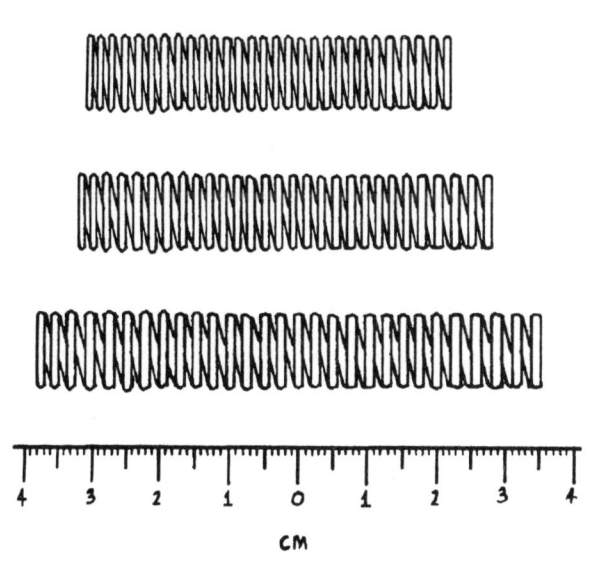

Fig 6.10 The three lengths of SMA spiral used by Qui *et al.*, from top: 50, 55, 65 mm (46)

Table 6.6 Clinical data of patients implanted by Soni *et al.* with the Memokath urethral stent (45)

Patient	Date of birth	Date of injury	Diagnosis	Past urological operation for urinary bladder emptying
1	19.04.56	18.01.76	C5 tetraplegia	TURBIN (24.01.78)
2	19.10.48	August 1967 (Viral infection)	D6 paraplegia	
3	10.04.64	08.07.88	C5 tetraplegia	
4	09.02.76	12.09.91	C6 tetraplegia	
5	07.03.47	12.04.84	C4 tetraplegia	
6	23.01.35	23.12.91	C5 tetraplegia	
7	08.06.65	15.06.86	C5 tetraplegia	
8	26.04.38	03.04.90	C7 tetraplegia	
9	10.03.56	30.04.73	C5 tetraplegia	Sphincterotomy, 27.11.73
10	25.10.58	09.03.85	C6 tetraplegia	DES, 10.10.86

TURBIN = transurethral resection of bladder neck.
DES = division of external urethral sphincter.

the placement site by the injection of 40 to 50°C water (a variation on the hot saline technique used in cardiovascular stenting), and Qui *et al.* noted that all stents were positioned correctly.

The group also found that

> The symptom scores of all patients had improved remarkably.

The group concluded that the NiTi material

> used in this study is an ideal biomaterial: we deem the spiral a good alternative in patients unfit for surgery.

6.3 Ni–Ti clinical instruments

The Randin report (**47**), along with other biocompatibility studies previously highlighted, have led some groups to investigate the viability of the Ni–Ti alloy for uses in clinical instruments.

Civjan *et al.* (**48**) reported on the use of tools fabricated from the alloy in 1975. The group looked into the properties of chisels made out of 60-nitinol, but went no further than discussing that:

> The characteristics of 60-nitinol suggests its use in fabrication of tough 'non-dulling' corrosion resistant hand or rotary cutting instructions or files.

Current bladder management	Date urethral stent inserted	Size of stent	Method of insertion
Intermittent catheterization by his mother	10.11.93	70 mm	Under fluoroscopic control
Self catheterization	21.09.93	60 mm	Under fluoroscopic control
Indwelling urethral catheter drainage	01.09.93	60 mm	Under fluoroscopic control
Indwelling urethral catheter drainage	03.08.93	50 mm	Under fluoroscopic control
Indwelling urethral catheter drainage	27.07.93	60 mm	Under direct vision with flexible cytoscope
Indwelling urethral catheter drainage	22.09.93	70 mm	Under fluoroscopic control
Indwelling urethral catheter drainage	17.11.93	60 mm	Under fluoroscopic control
Indwelling urethral catheter drainage	26.10.93	70 mm	Under direct vision with flexible cytoscope
Indwelling urethral catheter drainage	15.10.93	60 mm	Under direct vision with flexible cytoscope
Indwelling urethral catheter drainage	19.10.93	70 mm	Under fluoroscopic control

A more extensive study into one of these areas was produced by Walia et al. (**49**) in 1988. This group investigated the possible use of the Ni–Ti alloy as a root canal file.

In the instrumentation of curved canals certain difficulties can occur, these difficulties mainly being caused by the stiffness of the stainless steel alloy file used. Walia et al. compared a 0.2 inch diameter, 2 inch long Ni–Ti file with a stainless steel file of similar dimensions and concluded that:

> **The nitinol files were observed to have two or three times the elastic flexibility of stainless steel as well as a superior resistance to fracture.**

Walia et al. further concluded that:

> **Nitinol endodontic files may have particular promise for clinical preparation of curved root canals.**

Cuschieri (**50**) discussed the need for minimal access surgery to be performed for laparoscopic operations. Minimal access surgery (MAS) allows a reduction of access trauma, encourages speedy rehabilitation and reduces the risk of wound infection, compared to conventional surgery, and thus is of great benefit to patient and surgeon.

Cuschieri commented on the requirement for MAS of curved instruments that would be able to access the peritoneal cavity. In another report published

Fig 6.11 Laparoscopic cannular used by Cushieri *et al.* (51)

by Cuschieri in 1991 **(51)**, the author discussed the possible use of a Ni–Ti shape memory tool in relation to such an application.

Cuschieri reported on the application of the SMA in the context of a laparoscopic spatula that could achieve variable curvatures. The SMA used was memorized into a semicircle with a diameter of 2.5 cm. The instrument was then restrained within a standard laparascopic cannular (Fig. 6.11) and introduced into the peritoneal cavity. Once in the correct region the spatula was extended out of its cannular until the required curvature was achieved (Fig. 6.12(a)–(d)).

Cuschieri concluded that:

> The use of the superelastic curved spatula has considerably facilitated dissection of the cystic artery, cystic duct, and common bile duct.

The report continued that the SMA spatula:

> overcomes to a large extent these intrinsic restrictions of laparoscopic surgery.
>
> Our initial experience with the pseudoelastic curved spatula has been entirely favourable and indicates that laparoscopic instrumentation based on the mechanical shape memory phenomena of pseudoelastic alloys is needed for the advancement of the scope of endoscopic surgery.

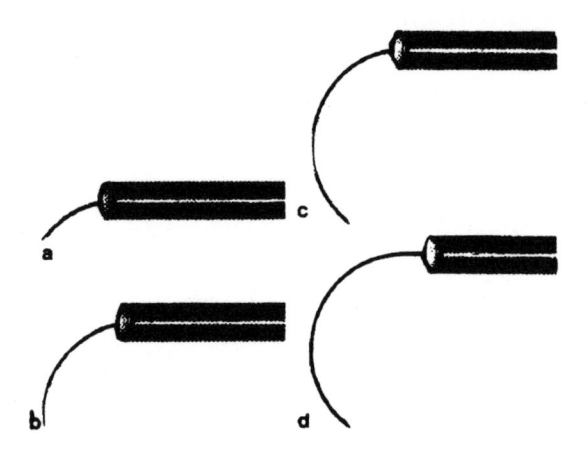

Fig 6.12 Example of variable curvature available from the Cushieri *et al.*'s SMA spatula (51)

Other Applications of the SMA

In 1990 Constantin Cope (**52**) published a paper using a new design of nitinol basket for capture and removal of calculi from hollow organs, such as the gall bladder. The nitinol basket was found to have several advantages over other captivity techniques available, and led Constantin to conclude:

The combination of a nitinol basket with a large instrument channel for irrigation, lithotripsy and endoscopy is a substantial technical improvement for rapidly breaking up and retrieving stones from the gall bladder.

In recent years the diagnosis and management of bilopancreatic diseases has been greatly advanced using a technique called peroral cholangiopancreatoscopy (PCPS). This technique involves the passing of a small calibre 'baby' scope through a larger 'mother' scope catheter and thus allowing direct visualization of the subject area.

However, these ultra thin scopes are not directional and the conventional mother scopes require endoscopic sphincterotomy (EST) to achieve entry into the duct system.

In 1994 Mizuno *et al*. (**53**) reported on the use of uni and bi directional NiTi SMA tipped mother catheters (Table 6.7) for use in PCPS.

Mizuno *et al*. applied the SMA system to nineteen patients and the results are shown in Table 6.8.

Table 6.7 Specifications of the shape memory alloy catheters

	Two-directional type	One-directional type
Whole length	2245 mm	2245 mm
External diameter	2.6 mm	2.2 mm
Internal diameter	1.2 × 2.2 mm	1.2 mm
Length of bending tip	20 mm	20 mm
Tip deflection	90° up/down	90° up

Table 6.8 Results of gain by Mizuno *et al*. using a SMA catheter system (53)

SMA catheter type	No. of patients	No. of successful cannulations papilla pancreas	No. of successful inspections
Two-directional	7	6 (5) 5	5
One-directional	12	12 (1) 12	11
Total	19	18† (6) 17‡	16*

Note: Numbers in brackets are the number requiring prior EST.
†One two-directional failed to enter papilla because the catheter tip was too soft.
‡Prevention of two-directional advancement severe tortuosity of main duct.
*Visualization failure as 'baby' scope could not be inserted due to the tortuosity of pancreatic duct.

Mizuno *et al*. concluded that altho[...] recommended:

> In our experience, the two pro[...] instruments for introducing an ultrat[...] system....
>
> ... Further refinements of the instruments would help improve our techniques in PCPS and enhance our abilities in the diagnosis and management of bilopancreatic diseases.

Reports on the retrieval of foreign bodies or misplaced devices have been made since 1954 (**54**). However, until the mid-60s the only method of removal was by surgery. Thomas *et al*. (**55**) changed that by publishing the first report of percutaneous removal in 1964 and today the non-surgical retrieval of unwanted intravascular bodies (Figs 6.13 and 6.14) has become standard (**56**) (**57**). Table 6.9 shows the range of foreign bodies removed by non-surgical techniques.

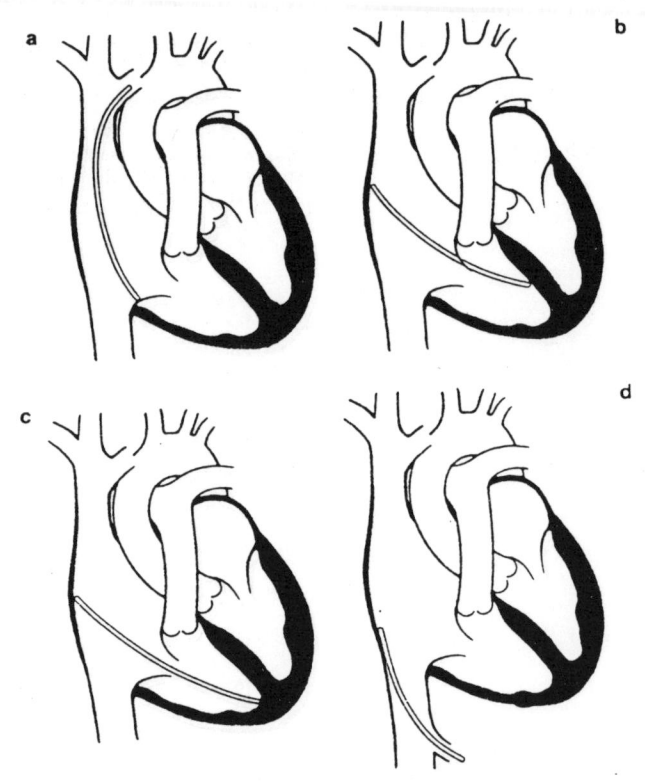

Fig 6.13 Various locations of foreign body migration shown with broken catheter example as discussed by **Gerlock** *et al*. (56)

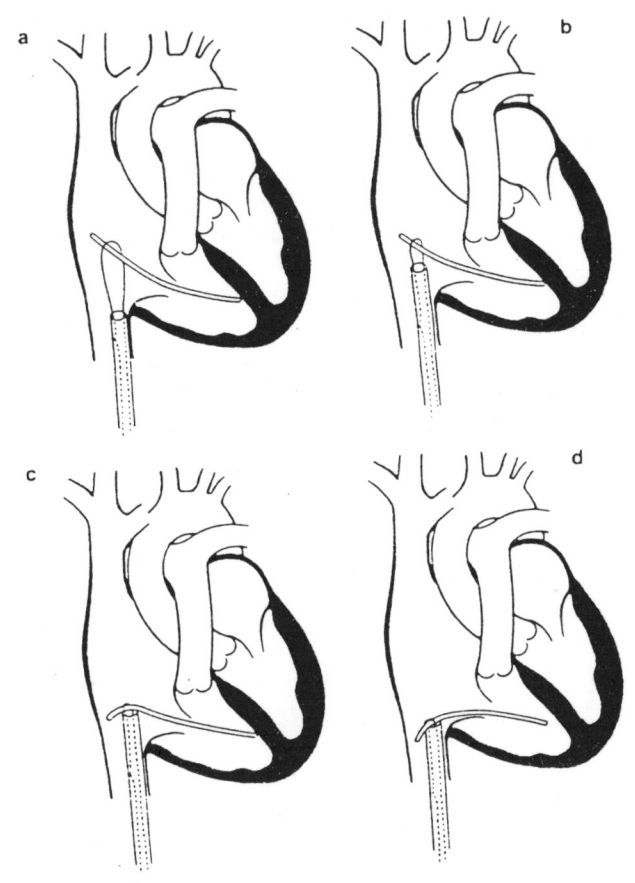

Fig 6.14 Principal stages of the snare loop technique (Gerlock *et al.* (56))

Gerlock *et al.* (**56**) produced a report in 1987 that discussed the main methods of internal foreign body removal. The group noted that the loop snare which, due to its simplicity and relative ease of application, had become the most widely used retrieval method (Fig. 6.15).

The 90° gooseneck snare (Fig. 6.15(b)) is an example of a loop retrieval system and is used when straight loops snaring is constrained.

Table 6.9 Foreign bodies removed from the vascular system as highlighted by Gerlock *et al.* (56)

Metal guide wire fragments	Catheter fragments
Gianturco coils	Central venous lines
Ventriculovenous shunts	Swan–Ganz catheter
Pacing wire	Holter valve
Angiographic dilators	

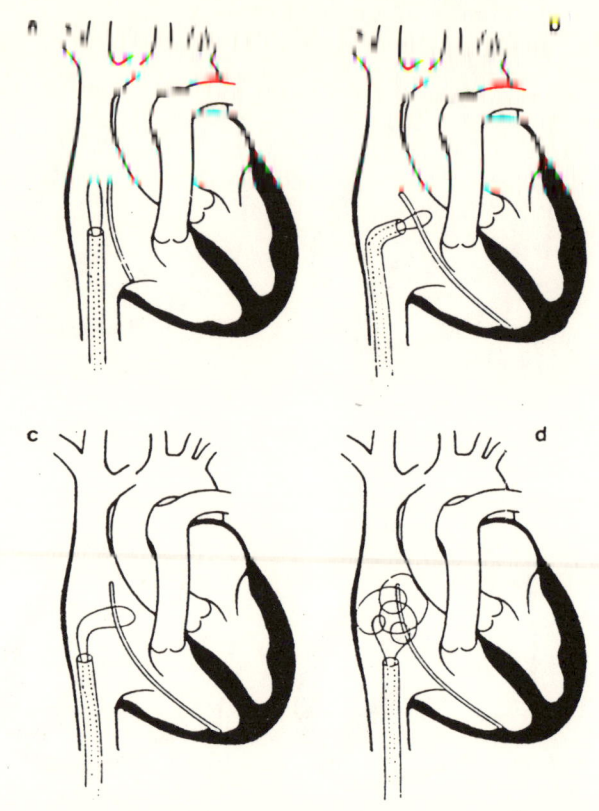

Fig 6.15 Methods used to retrieve fragment when its long axis is parallel to the introducing catheter (Gerlock *et al.* (56)). (a) Long axis of body almost parallel with introducing catheter, straight snare is unable to remove body; (b) Curved tipped wire allows loop snare to encircle fragment; (c) Lateral bend in the wire (90 degree snare) works in a similar fashion to (b); (d) Multiple loop solution.

However, until recently problems had arisen with its use due to the basic physical properties of the constituent stainless steel. In 1991 Yedlicka *et al.* (**58**) published a report on the experimental study in the use of an Ni–Ti snare (Fig. 6.16), to remove implanted bodies in four mongrel dogs (20–25 kg).

The investigation also reported four internal stent removals and three intravascular foreign body retrievals. Yedlicka *et al.* found that times for the operation were short, on average 15 minutes, and discussed the comparative ease and high success rate of retrieval shown by the Ni–Ti snare with respect of an equivalent snare made from stainless steel. Yedlicka *et al.* concluded:

> This versatile snare and catheter system has proved very easy to use ... We prefer the gooseneck nitinol snare in all retrieval situations.

In 1994 Cerkirge *et al.* (**59**) used four sizes of gooseneck snare to remove twenty varying foreign bodies from the vascular system ($n = 13$), urinary

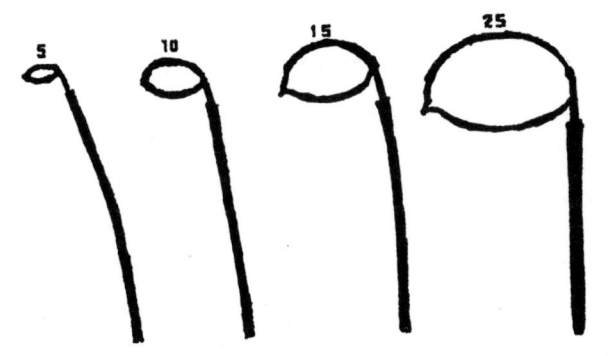

Fig 6.16 The Gooseneck snare as implemented by Yedlicka et al. (58)

system ($n = 4$), biliary system ($n = 1$), gastrointestinal tracts ($n = 1$), and the peritoneal space ($n = 1$) (see Table 6.10).

Cerkirge et al. found:

Removal of the retained foreign body was successful in all patients.

Table 6.10 Summary of foreign bodies extracted by Cerkirge et al. (59)

Summary of type and location of retrieved foreign bodies

Patient No.	Foreign body	Location
1	Broken CVA catheter	Superior vena cava
2	Broken CVA catheter	Right atrium
3	Broken CVA catheter	Between the right hepatic & left pulmonary artery
4	Broken CVA catheter	Right atrium
5	Broken CVA catheter	Right internal jugular
6	Broken CVA catheter	Right atrium
7	Retained teflon sheath	Right femoral vein
8	Retained subclavian dialysis catheter	Inferior vena cava
9	Retained fragment of 0.035-inch guide wire	Inferior vena cava
10	42-mm Wallstent	Right atrium
11	3-mm embolization coil	Splenic vein
12	Retained fragment of a balloon catheter	Femoropopliteal venous bypass graft
13	Retained fragment of a balloon catheter	Renal artery
14	Retained fragment of 0.018-inch guide wire	Common bile duct
15	Occluded ureteral stent	Ureter
16	Occluded ureteral stent	Ureter
17	Three pieces of a broken ureteral stent	Renal pelvis and proximal ureter
18	Five pieces of a broken ureteral stent fragment	Renal pelvis and proximal ureter
19	Misplaced double pigtail catheter	Stomach
20	18 × 18-inch surgical laparotomy sponge	Peritoneal cavity

They also reported that:

There were no procedural or post-procedural complications in any of these cases.

The group concluded:

The nitinol gooseneck snare has proved easy to use in a wide variety of clinical settings. Our technical success and lack of complications have led as to adopt this device as first-time therapy for all foreign body removals.

In a further three reports written in 1994, Patel *et al.* (**60**), Sommer *et al.* (**61**) and Dixon (**62**) utilized the nitinol gooseneck snare for the repositioning of a ureteric stent, the aiding of a trancatheter coil delivery and the removal of abscess debris respectively. All three groups reported that the snare was successful in the application for which it was required and gave encouraging implications for future use.

It can be seen that development over the last thirty years in Ni–Ti instruments has been slow and has mainly concentrated on the elasticity rather than the shape memory property of the alloy. Further advancements in Ni–Ti tools are bound to occur and the implications look very promising for their use in many fields.

6.4 Cranial aneurysm clips

An aneurysm is the formation of a sac of sometimes pulsating blood caused by the dilation or weakening of a vessel wall. If such an upper body aneurysm is formed in the cranial region, then it can be extremely serious; aortic aneurysms alone have a mortality rate of 80 percent even after emergency surgery (**63**).

Therefore, the fast and effective treatment of such conditions is essential for the well being of the subject. To overcome cranial aneurysms modern techniques use either a V-shaped clip to halt blood flow between the vessel and the sac and, by cutting the blood flow, the aneurysm may be gradually healed; or by using stents to acclude the aneurysm opening. However, due to the nature of the operation, this surgery is very delicate and mechanical or human failure may create surrounding tissue damage and risk to life.

Honma *et al.* (**64**) used the Ni–Ti alloy in four different types of cranial clips:

(a) V-shaped
(b) pin-shaped
(c) Yosargil type
(d) connecting type.

Through investigation Honma *et al.* found that all of the first three types showed problems in their design. The connecting type design (Fig. 6.17), however, showed promise. Its design was a combination of silver and Ni–Ti

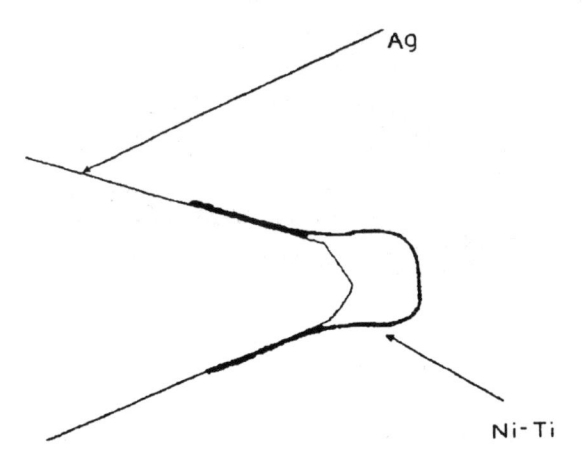

Fig 6.17 Ag–Ni–Ti connecting type design by Honma *et al.* (64)

and, although the connection between the two metals was difficult to achieve satisfactorily, it was clearly superior to other types that were fabricated.

Honma *et al.* concluded that further improvements were required before the design could be clinically applied.

In 1993 Hagen *et al.* (**65**) reported on the use of two types of nitinol stents in the aim of excluding artificial aortic aneurysms in 11 fox hounds. The group used two types of prosthesis; single mesh ($n = 2$) and double mesh ($n = 8$) (Fig. 6.18) and compared the results. One canine was lost due to a suture infection prior to stent implantation.

The single mesh and double mesh designs were each constructed from 0.1 mm diameter wire and had sixteen loops per circumference. However, the single and double designs differed slightly in pore sizes between loops 0.84 mm^2 and 0.68 mm^2, respectively.

Hagen *et al.* misplaced one prosthesis which thus failed to span the entire aneurysm. Of the other nine, the seven double meshed stents showed immediate, complete exclusion, and the single meshed stents showed only partial or delayed occlusion. No complications or long term reduction in lumen patency was reported, and a neointimal layer had formed over the whole stent at one year. The group concluded that:

> Our preliminary results indicate an effective exclusion of artificial aortic aneurysms in dogs by tightly knitted porous stent grafts.

Another publication in 1994 by Wakhloo *et al.* (**66**) looked into the application of very similar single mesh nitinol stents on the exclusion of carotid aneurysms.

Wakhloo *et al.* implanted fifteen stents into labrador carotid arteries with artificial aneurysms, and the effectiveness of the nitinol stent ($n = 10$) (Fig.

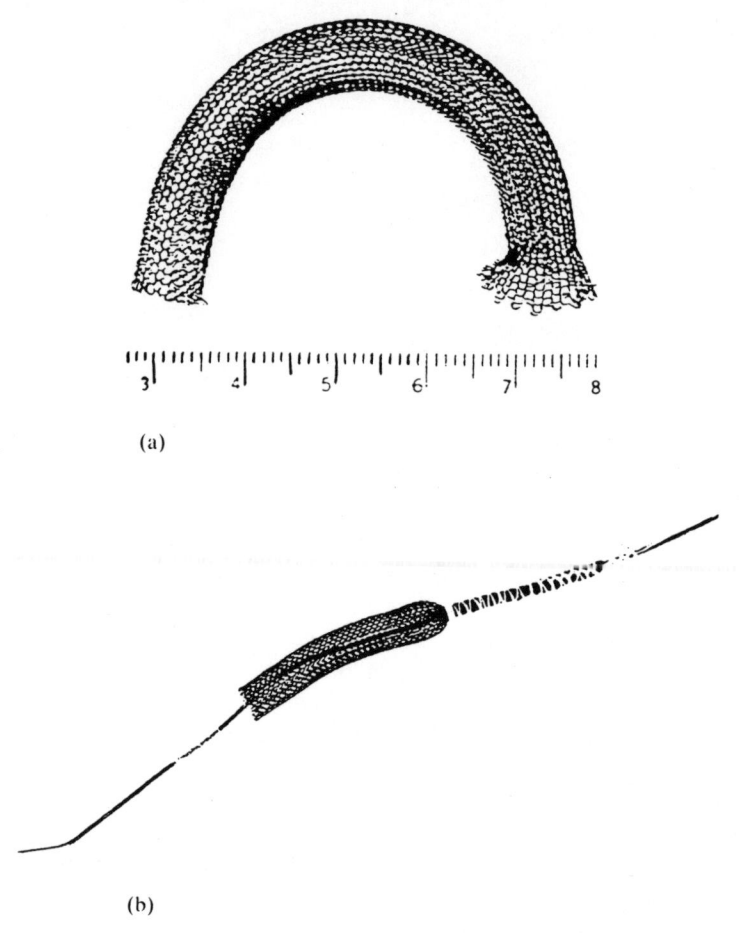

Fig 6.18 (a) Double knitted macroporous stent. (b) Delivery system used by Hagen et al. (65)

6.19), and a balloon expandable tanatalum stent ($n = 5$) were assessed and compared.

The group reported that apart from in two tantalum stent cases, all the prostheses produced either immediate occlusion of the aneurysm ($n = 7$), or delayed complete thrombosis after 7–10 days ($n = 3$). In two cases stenosis of the carotid lumen occurred; tantalum ($n = 1$), nitinol ($n = 1$) and in a single case two nitinol stents had to be placed in tandem (overlapping) to cover the size of the aneurysm.

Wakhloo et al. found that the nitinol endoprosthesis had considerably lower amounts of stenosis than that of a balloon expandable tantalum implant (Fig. 6.20), and the group concluded that:

> Intravascular stents are potential treatment for aneurysm, especially large,

Other Applications of the SMA 131

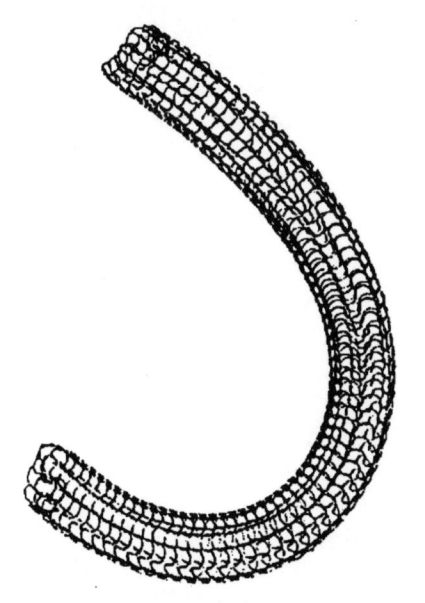

Fig 6.19 Fully expanded NiTi strecker stent used by Wakhloo *et al.* (66). (Copyright (1994) American Society of Neuroradiology.)

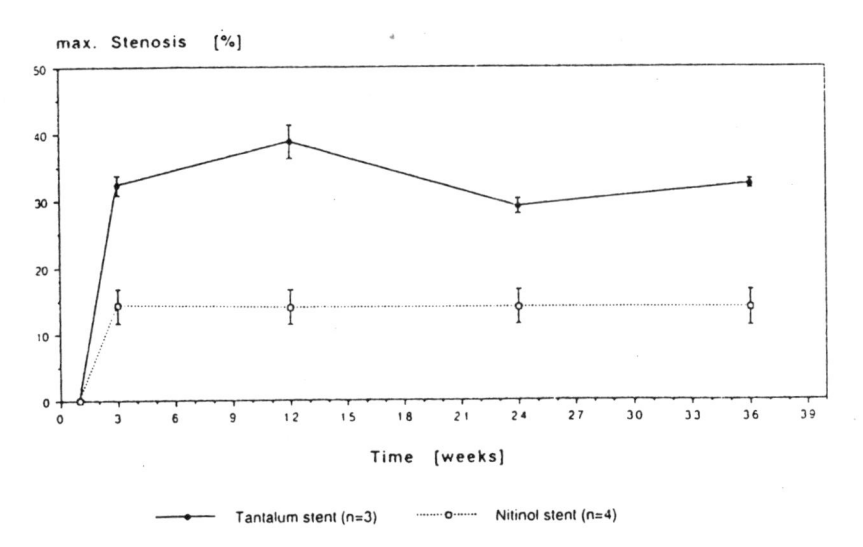

Fig 6.20 Angliographic medium-term follow up of stents implanted by Wakhloo *et al.* (66). (Copyright (1994) American Society of Neuroradiology.)

fusiform aneurysms. They may result in less danger to the parent vessel and less risk of rupturing the aneurysm. However, more detailed knowledge of the hemo-dynamics of the aneurysm, including the total volume and the diameter of the neck, is required before a suitable type of stent can be designed.

6.5 Intravascular grafts

When the thoracic or abdominal aorta requires replacing, a sutureless intraluminal vascular graft is now the accepted means to manage the requirements needed. However, such prostheses suffer from complications caused by their inflexibility during implantation, their non-rigidity on final placement and the degeneration and positional difficulties caused by the need for securing ties.

In 1994 Yang *et al*. (**67**) reported on a new sutureless intraluminal (Dacron) graft made with SMA NiTi sealing rings (Fig. 6.21).

Yang *et al*.'s prosthesis allowed flexible implantation below 38°C and became rigid once flushed with warm saline.

The group implanted nine prostheses into the descending thoracic aorta of nine mongrel dogs. Follow up occurred between 15 and 90 days (Table 6.11), and no graft related deaths or anastomostic bleeding was observed.

Yang *et al*. were very encouraged by their initial report and commented that:

> Modification of the sutureless prosthesis with titanium nickel shape memory alloy rings will suit the pathologic conditions encountered at operation and allow the quickest repair with the least chance of anastomotic complication.
>
> The new graft is suitable for aortic and peripheral arterial reconstruction.

Also in 1994 Yoshida *et al*. (**68**) described their initial development of an insertable aortic prosthesis. Yoshida's *et al*. advice consisted of a loose round nitinol stent (transition temperature 30°C) with a polyurethene tube covering.

The group evaluated the prosthesis in three major steps:

(a) preliminary canine endoprosthetic trials
(b) percutaneous canine and pig endoprosthetic trials
(c) application to a dissection model.

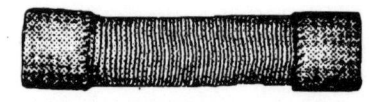

Fig 6.21 The sutureless intraluminal graft used by Yang *et al*. (67)
Note: (a) the ring is made up of flat SMA with a 15 mm diameter;
(b) the ring can be compressed to 8 mm in diameter in 4°C normal saline solution;
(c) the ring can recover its original shape after 20 ml normal saline solution at 38°C is injected onto its surface.

Other Applications of the SMA

Table 6.11 Summary of Yang et al.'s (67) crossclamp and corresponding follow up periods

Dog No.	Crossclamp time (min)	Follow up period (days)
1	4	15
2	4	23
3	2	32
4	3	39
5	2	48
6	3	57
7	3	65
8	2	76
9	3	90

In the initial step, stents were deployed into ten canine subjects, with access being gained from the canine common carotid (catcher catheter) and femoral (delivery catheter) arteries (Fig. 6.22).

Three sizes of prosthesis were introduced, and by comparing the histological effects and migrational occurrences (Table 6.12), the group hypothesized the optimum shape memory aortic prosthesis (SAP) diameter. The group implemented this 'optimum' prosthesis diameter (group 3) in the subsequent investigations.

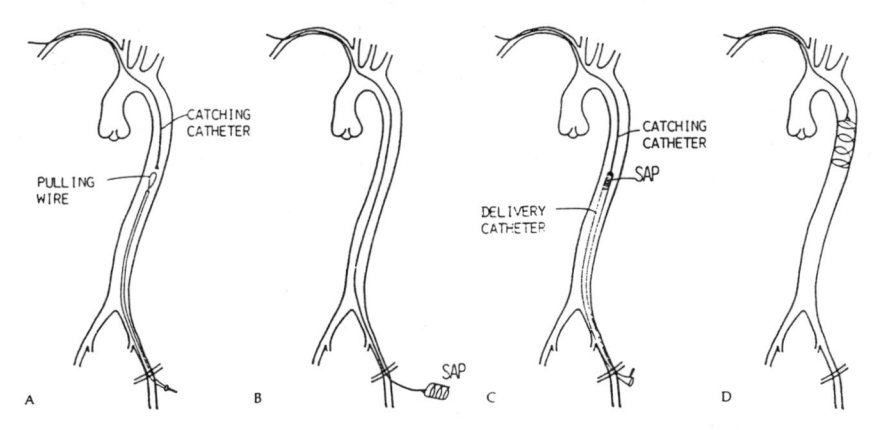

Fig 6.22 Transcatheter placement of the Yoshida et al. SMA Graft (68)

Table 6.12 Generated groups used by Yoshida et al. (1994) (68)

Group No.	Prosthetic diameter (s) (aortic diameter = A)	No. of dogs	No. of migration	Histological damage
Group 1	s = A − 1 mm	2	2	None
Group 2	s = A + 1 mm	2	2	None
Group 3	s = A + 3 mm	6	0	Minimal

The group moved on to percutaneously implant four SAPs into four animal subjects (three dogs and one sheep) and Yoshida *et al.* observed that all placements went without incidence.

Finally, and as the 'most important step' of the study, Yoshida *et al.* transfemorally implanted three SAPs into canines, in which a 'standard type B' aortic dissection model had been created. One dog was lost to follow up due to aortic rupture caused by a prosthesis placement problem, but the other two canines appeared healthy in follow up angiograms after three months. Yoshida *et al.* concluded that:

> **The SAP shows potential for use in treating emergency cases of aortic rupture, closure of entry points for standard type B aortic dissections and high risk cases of aortic aneurysms.**

6.6 Conclusion

This chapter covers a wide range of topics from SMA dental arch wires, which are presently implemented by orthodontic specialists, to thermal airway stents, various laparoscopic techniques, and foreign body removal snares.

In all the cases discussed within this chapter, the advantage of the Ni–Ti SMA has been made clear, from their lowering of discomfort in orthodontic patients to the increasing of a 90° snare's working range.

Orthodontic arch wire application has already reached the clinical stage with large advantages being concluded from its initial reports. All other applications within this chapter have not advanced so far, and are still at the development stage.

This chapter shows clearly the diversity possible and gives an insight into how broad a field of interest there is concerning Ni–Ti SMA application. All cases have shown good responses to the alloy and go some way to proving that its unique ability is perfect for a wide range of operations with advantage over modern conventional techniques.

References

(1) BURSTONE, C. J., and GOLDBERG, A. J. Beta titanium, a new orthodontic alloy, *Am J. Orthod*, 1980, **77**, 121–132.

(2) KUSY, R. P., and GREENBERG, A. R. Effects of composition and cross section on the elastic properties of orthodontic wires, *Angle Orthod*, 1981, **51**, 325–341.

(3) GRABER, T. M. *Orthodontics: Principles and Practice*, 1972 (W. B. Saunders Co, Philadelphia), 556–560.
(4) KOHL, R. W. Metallurgy in orthodontics, *Angle Orthod*, 1964, **34**, 37–52.
(5) STRANG, R. S. W., and THOMPSON, W. M. *A text book of orthodontics*, 1958 (Lea and Febigec), 598.
(6) THURAU, R. C. *Edgewise Orthodontics* (C. V. Mosby Co, St Louis, Missouri).
(7) BEGG, P. R. Differential forces in orthodontic treatment, *Am J. Orthod*, 1956, **42**, 481–510.
(8) STONER, M. M. Force control in clinical practice, *Am J. Orthod*, 1960, **46**, 163–186.
(9) JARABAK, J. R. Development of a treatment plan in the light of one's concept of treatment objectives, *Am J. Orthod*, 1960, **46**, 481–514.
(10) BURSTONE, C. J., BALDWIN, J. J., and LAWLESS, P. T. The application of continuous forces to orthodontics, *Angle Orthod*, 1961, **31**, 1–14.
(11) WATERS, N. E., HOUSTON, W. J. B., and STEPHENS, C. D. The characterization of arch wires for initial alignment of irregular teeth, *Am J. Orthod*, 1981, **79**, 373–389.
(12) SIMS, M. R. Loop systems – a contemporary reassessment, *Am J. Orthod*, 1972, **61**, 271–278.
(13) ANDREASEN, G. F., and BRADY, P. R. The use hypothesis for 55 nitinol wire for orthodontics, *Angle Orthod*, 1972, **42(2)**, 172–177.
(14) ANDREASEN, G. F., BIGELOW, H., and ANDREWS, J. G. 55 nitinol wire: force developed as a function of 'elastic memory', *Aust Dent J.*, 1979, **24**, 146–149.
(15) ANDREASEN, G. F. A clinical trial of alignment of teeth using a 0.019 inch thermal nitinol wire with a transition temperature range between 31°C and 45°C, *Am J. Orthod*, 1980, **78(5)**, 528–537.
(16) ANGLE, E. H. Classification of malocclusion, *Dent Cosmos*, 1899, **41**, 248–264.
(17) FOSTER, T. D. A textbook of orthodontics, 1990, 3rd Ed (Blackwell Scientific Pub).
(18) PETERSON, L., SPENCER, R., and ANDREASEN, G. F. A comparison of friction resistance for nitinol and stainless steel wire in edgewise brackets, *Quintess Int*, May 1982, **5**, 563–571.
(19) ANDREASEN, G. F., MONTAGANO, L., and KRELL, D. An investigation of linear dimension changes as a function of temperature in an 0.010 inch 55 cobalt-substituted annealled nitinol alloy wire, *Am J. Orthod*, 1982, **82(6)**, 469–472.
(20) BURSTONE, C. J., QIN, B., and MORTON, J. Y. Chinese Ni–Ti wire – a new orthodontic alloy, *Am J. Orthod*, 1985, **87(6)**, 445–452.
(21) MIURA, F., MOGI, M., OHURA, Y., and HAMANAKA, H. The

super-elastic property of the Japanese Ni–Ti alloy wire for use in orthodontics, *Am J. Orthod*, 1986, **90(1)**, 1–10.
(22) TIDY, D. C. Frictional forces in fixed applications, *D Orth Am J. Orthod Denofac Orthop*, 1989, **96(3)**, 249–254.
(23) DRESCHER, D., BOURAUEL, C., and SCHUMACHER, H. A. Frictional forces between bracket and arch wire, *Am J. Orthod Dentofac Orthop*, 1989, **96(5)**, 397–404.
(24) HURST, C. L., DUNCANSON, M. G., NANDA, R. S., and ANGOLKAR, P. V. An evaluation of the shape memory phenomenon of nickel titanium orthodontic wires, *Am J. Orthod Dentofac Orthop*, 1990, **98(1)**, 72–76.
(25) KAPILA, S., REICHOLD, G. W., ANDERSON, R. S., and WATANA, L. G. Effects of clinical recycling on mechanical properties of nickel–titanium alloy wires, *Am J. Orthod Dentofac Orthop*, 1991, **100(5)**, 428–435.
(26) MOHLIN, B., MULLER, H., ODMAN, J., and THILANDER, B. Examination of Chinese Ni–Ti wire by a combined clinical and laboratory approach, *Euro J. Orthod*, 1991, **13**, 386–391.
(27) KAPILA, S., HAUGEN, J. W., and WATANABE, L. G. Load-deflection characteristics of nickel titanium alloy wire after clinical recycling and dry heat sterilization, *Am J. Orthod Dentofac Orthop*, 1992, **102(2)**, 120–126.
(28) ANDREASEN, G. F., and AMBORN, R. M. Aligning, levelling and torque control – a pilot study, *Angle Orthod*, 1989, **59(1)**, 51–60.
(29) SHIM, C. S., LEE, M. S., KIM, J. H., and CHO, S. W. Endoscopic applications of Gianturco-Rosch biliary z-stent, *Endoscopy*, 1992, **24**, 436–439.
(30) BETHGE, N., KNYRIM, K., and WAGNER, H. J. *et al*. Self expanding metal stents for palliation of malignant esophageal obstruction – a pilot study of 8 patients, *Endoscopy*, 1992, **24**, 411–415.
(31) VAN ARSDALEN, K. N., POLLACK, H. M., and WEIN, A. J. Ureteral Stenting, *Seminars in Urology*, 1984, **2(3)**, 180–186.
(32) SIMONDS, A. K., IRVING, J. D., CLARKE, S. W., and DICK, R. Use of expandable metal stents in the treatment of bronchial obstructions, *Thorax*, 1989, **44**, 680–681.
(33) WALLACE, M. J., CHARNSAGAVEJ, C., and OGAWA, K. *et al*. Tracheo-bronchial tree: expandable metallic stents use in experimental and clinical applications, *Radiology*, 1986, **158**, 309–312.
(34) GEORGE, P. J., IRVING, J. D., MANTELL, B. S., and RUDD, R. M. Covered expandable metal stent for recurrent tracheal obstruction, *Lancet*, 1990, **335**, 582–584.
(35) NEVILLE, W. E., BOLANOWSKI, P. J. P., and SOLTANZADEH, H. Prosthetic reconstruction of the trachea and carina, *J. Thorac Cardiovasc Surg*, 1976, **72**, 525–538.

(36) RAUBER, K., FRANKE, C., RAU, W. S., SYED ALI, S., and BENSMANN, G. Perorally insertable endotracheal stents made from Ni–Ti memory alloy – an experimental animal study, *(German) Refo: Fortschritte Auf Dem Gebiete der Rontgenstrahlen Und Der Nuklearmedizin*, 1990, **152(6)**, 698–701.
(37) NAKAMURA, T., SHIMIZU, Y., and MATSUI, T. *et al*. Novel airway stent using a thermal shape-memory Ti–Ni alloy, *ASAIO Trans*, 1991, **37(3)**, M319–M321.
(38) HANAWA, T., IKEDA, S., and FUNATSU, T. *et al*. An experimental study on surgical treatment for tracheomalacia: Methods for making models of tracheomalacia and development of a new operative method, *Nippon Geka Gakkai Zasshi*, 1989, **90**, 1072–1080.
(39) NAKAMURA, T., SHIMIZU, Y., and ITO, Y. *et al*. A new thermal shape memory Ti–Ni alloy stent covered with silicone, *ASAIO Trans*, 1992, **38**, M347–M350.
(40) VINOGRAD, I., KLIN, B., and BROSH, T. *et al*. A new intratracheal stent made from nitinol, an alloy with shape memory effect, *J. Thorac Cardiovasc Surg*, 1994, **107**, 1255–1261.
(41) DOTTER, C. T. Transluminally placed coil-spring end arterial tube grats, long term patency in canine popliteal artery: Investigation, *Radiology*, 1969, **4**, 329.
(42) MILROY, E. J., CHAPPLE, C. R., and COOPER, C. *et al*. A new stent for the treatment of urethral strictures, *Lancet*, 1988, **1**, 14–24.
(43) YACHIA, D. The use of urethral stents for the treatment of intraprostatic spirals, *Ann Urol*, 1993, **27(4)**, 245–252.
(44) POULSEN, A. L., SCHOU, J., and OVERSEN, H. *et al*. Memokath: a second generation of intraprostatic spirals, *Br J. Urol*, 1993, **72**, 331–334.
(45) SONI, B. M., VAIDYANATHAM, S., and KRISHNAN, K. R. Use of memokath, a second generation urethral stent for relief of urinary retention in male spinal cord injured patients, *Paraplegia*, 1994, **32**, 480–488.
(46) QUI, C. Y., WANG, J. M., and ZHANG, Z. X. *et al*. Stent of shape memory alloy for urethral obstruction caused by benign prostatic hyperplasia, *J. Endourol*, 1994, **8(1)**, 65–67.
(47) RANDIN, J. P. Corrosion behaviour of nickel containing alloys in artificial sweat, *J. Biomed Mater Res*, 1988, **22**, 649–666.
(48) CIVJAN, S., HUGET, E. F., and DeSIMON, L. B. Potential applications of certain nickel–titanium (nitinol) alloys, *J. Dent Res*, 1975, **54(1)**, 89–96.
(49) WALIA, H., BRANTLEY, W. A., and GERSTEIN, H. An initial investigation of bending and torsional properties of nitinol root canal files, *J. Endodontics*, 1988, **14(7)**, 346–351.
(50) CUSCHIERI, A. Minimal access surgery and the future of interventional laparoscopy, *Am J. Surg*, 1991, **161**, 404–408.

(51) CUSCHIERI, A. Variable curvature shape-memory spatular for laparoscopic surgery, *Surg Endosc*, 1991, **5**, 179–181.
(52) COPE, C. Novel nitinol basket instrument for percutaneous cholecystolithotomy, *AJR*, 1990, **155**, 515–516.
(53) MIZUNO, S., NAKAJIMA, M., and YASUDA, K. *et al.* Shape memory alloy catheter system for peroral pancreatoscopy using a ultrathin-caliber endoscope, *Endoscopy*, 1994, **26**, 676–680.
(54) TURNER, D. C., and SOMMERS, S. C. Accidental passage of a polyethylene catheter from a cubital vein to right atrium: fatal case, *New Eng J. Med*, 1954, **251**, 744–745.
(55) THOMAS, J. S., SINCLAIR-SMITH, B., and BLOOMFIELD, D. *et al.* Non-surgical retrieval of a broken segment of steel spring guide from right atrium and inferior vena cava, *Circulation*, **30**, 106–108.
(56) GERLOCK, A. J., and MIRFAKHRASE, M. Retrieval of intravascular foreign bodies, *J. Thorac Imag*, 1987, **2(2)**, 52–60.
(57) UFLACKER, R., LIMA, S., and MELICHAR, A. C. Intravascular foreign bodies: percutaneous retrieval, *Radiology*, 1986, **160**, 731–735.
(58) YEDLICKA, J. W., CARBON, J. E., and HUNTER, D. W. *et al.* Nitinol gooseneck snare for removal of foreign bodies: experimental study and clinical evaluations, *Radiology*, 1991, **178**, 691–693.
(59) CERKIRGE, S., WEISS, J. P., and FOSTER, R. G. *et al.* Percutaneous retrieval of foreign bodies: Experience with the nitinol gooseneck snare, *JVIR*, 1993, **4**, 805–810.
(60) PATEL, U., and KELLETT, M. J. The misplaced double J ureteric stent: Technique for repositioning using the nitinol 'gooseneck' snare, *Clin Radiol*, 1994, **49**, 333–336.
(61) SOMMER, R. J., GUTIERREZ, A., and LAI, W. W. Use of preformed nitinol snare to improve transcatheter coil delivery in occlusion of patent ductus arteriosus, *Am J. Cardiol*, 1994, **15**, 836–839.
(62) DIXON, D. G. Nitinol snare for removal of abscess debris, *JVIR*, 1994, **5**, 647–648.
(63) PRINCE, M. R., SALZMAN, E. W., and SCHOEN, F. J. *et al.* Local intravascular effects of the nitinol wire blood clot filter, *Invest Radiol*, 1988, **23**, 294–300.
(64) HONMA, T., IWABUCHI, T., and NETSU, N. *Bull Res Inst Min Press Met Tohuku Univ*, 1978, **34(1)**, 67–73 (Japanese).
(65) HAGEN, B., HARNOSS, B. M., and TRABHARDT, S. *et al.* Self-expandable macroporous nitinol stents for transfemoral exclusion of aortic aneurysms in dogs: Preliminary results, *Cardiovasc Intervent Radiol*, 1993, **16**, 339–342.
(66) WAKHLOO, A. K., SCHELLHAMMER, F., and de VRIES, J. *et al.* Self-expanding and balloon-expanding stents in the treatment of carotid aneurysms: An experimental study in a canine model, *AJNR*, 1994, **15**, 493–502.

(67) YANG, C., SUN, Y., and DONG, P. *et al.* Experimental study of a new sutureless intraluminal graft with a shape-memory alloy ring, *J. Thorac Cardiovasc Surg*, 1994, **107**, 191–195.
(68) YOSHIDA, H., YASUDA, K., and TANABE, T. New approach to aortic dissection: Development of an insertable aortic prothesis, *Ann Thorac Surg*, 1994, **58**, 806–810.

CHAPTER 7

Conclusion

Since its discovery in 1963 by Buehler *et al.*, the shape memory phenomenon has been investigated on a vast array of fronts. Its physical and structural uniqueness has been discussed in many papers and the applications made possible by this phenomenon have been widely researched.

The shape memory phenomenon and the alloys that possess it, the shape memory alloys, were initially discovered in a near equiatomic alloy nickel and titanium by researchers at the US Naval Ordnance Laboratory, Maryland, USA. The alloy could, depending on consistency, show two distinct and, at the time, unique qualities:

(a) a superelastic response if work hardened, and
(b) a distinctive structural difference between low and high temperatures when quench cooled, which enabled shapes set at high temperatures to be reformed once the alloy was reheated, hence, a 'memory' ability displayed.

Since those early days, many other alloy systems have been discovered that exhibit the same 'memory' response as that of stochimetric Ni–Ti, each to a varying extent and differing percentage recovery.

In 1968 Buehler and Wang initialized the discussion of the possible engineering applications for the nickel–titanium alloy. However, it was not until 1972 that Andreasen and Brady postulated the first medical application of the shape memory alloy. The group discussed the feasibility of the alloy as a form of archwire, realigning malpositioned teeth, and concluded in a very favourable way.

So the concept had been started. Biocompatibility reports from groups such as Cutright (1973), Castleman (1976), Randin (1988), and Prince (1988) confirmed the biocompatibility of Ni–Ti to be of the same order as existing biomaterials, such as stainless steel and cobalt chromium alloy, and encouraged further research to be continued.

In 1977 Morris Simon *et al.* made the first study of a possible nickel–titanium shape memory alloy inferior vena cava filter, for the prevention of pulmonary embolisms within the human body. Simon *et al.*'s initial design was advanced by another study group lead by Palestrant *et al.* (1982), which also involved M. Simon, and the modern Simon nitinol filter was conceived.

In 1979 Schettler *et al.* initiated orthopaedic research in the shape memory phenomenon with tests on simulated jaw fractures. Schettler's work showed encouraging signs and, although slow to be investigated in the West, a lot of work has been produced on the SMAs possible application in China.

In 1983 Dotter *et al.* and Cragg *et al.* produced reports involving the first application of a shape memory alloy intravascular endoprosthesis or stent.

Since 1987 many further applications and investigations have been documented in fields ranging from intravascular foreign body removal apparatus to external muscular spacistity reducing splints. The area of interest concerning the nickel–titanium shape memory alloy is ever widening. Up to the time of this book, the commercial development and SMA studies have involved only three applications:

(a) The Simon nitinol filter – this was approved by the Food & Drinks Administration around the early 1990s and has become one of the leading new filters in the era of mechanical pulmonary embolism prevention.
(b) Arch wires – these have been clinically applied and have shown vast improvements on conventional dental procedures, cutting recovery times and allowing fewer follow up wire charges.
(c) Orthopaedic staples and arthroplastic cups – both clinically applied in the Far East, although not utilized by Western hospitals, have shown good results within the Chinese studies.

Other applications discussed within this book are developing at a rapid rate. The amount of papers produced each year on shape memory alloy application is ever increasing and the field of its utilization ever expanding.

This book discusses the applications to which the nickel titanium shape memory alloy has been investigated. It comments on the progress from initial historical background up to modern techniques. The shape memory phenomenon, its unique abilities and the vastness of its possible applications, make it a very worthy metal for further research. It has been shown to be as biocompatible as conventional biomaterials and has many major advantages.

The shape memory alloy is an alloy for the future. Although further tests need to be performed, it shows all possibility of becoming a major factor within the medical world.

Glossary

A

Abdominal – Pertaining to the front portion of the body between chest and pelvis.

Aggregation – A clumping of material.

Albers-Schönberg disease – Rare hereditary disease in which fractures occur easily.

Alveolar – Pertaining to cavities in either jaw, in which the roots of the teeth are imbedded.

Anastomosis – Cross connection of vessels.

Aneurysm – A sac formed by the dilation of a section of a vascular vessel.

Angiographic – Recording of vessel structure upon X-ray/gamma ray sensitive film.

Angioplasty – The act of repair on blood vessels.

Annealed – When a metal is heated up and allowed to cool slowly.

Anodic polarization – A method of corrosion resistance testing where a voltage is placed between a sample anode and a nominal cathode, which are in a physiological solution. The voltage is increased until the anode begins to ionize, causing a sharp rise in current. The higher the voltage the greater the resistance to corrosion.

Anteroposterior – The plane from the back of the body to front.

Anticoagulant therapy – Technique in which drugs are administered to prevent thrombotic formation.

Antiinflammatory – Any act to reduce inflammation and thus reduce risk of thrombosis.

Antiplatelet therapy – Administration of drugs to reduce formation of platelets.

Aorta – Artery originating from left ventricle and supplying the body with oxygenated blood.

Arthrodesis – Surgical fusion of a joint.

Artificial sweat – A solution produced to simulate the attributes of natural sweat.

Aseptic necrosis – A condition in which cell death occurs without inflammation.

Asymptomatic – The showing of no symptoms.

Austensite – A crystal structure comprising an ordered arrangement of atoms.

B
Balloon angioplasty – A method of angioplasty by which a balloon is inflated within the arterial passage so that the artery wall is expanded and the effective lumenal diameter increased.
Benign – Not malignant.
Bifurcation – The point at which division into two branches occurs.
BIH – Beth Israel Hospital, Harvard Medical School, Boston, MA.
Bile – The fluid secreted by the liver and poured into the duodenum as an aid to the digestive process.
Biliary – Pertaining to the bile ducts or gallbladder.

C
Canine – Pertaining to dogs; in experimental context, an investigation on dogs.
Capture rate – The success of an IVC filter to trap emboli within its frame.
Carotid artery – Principal artery of the neck.
Catheter – A flexible tubular instrument, passed through body channels for the removal or introduction of fluid/objects.
Cholangiopancreatography – X-ray examination of the bile ducts and pancreas after administration of a contrast medium.
Claudication – Lack of blood supply to the leg, causing limping.
Coagulability – The state of being capable of forming or being formed into clots.
Cochemia – Deficiency of blood in a part.
Collagenous tissue – A fibrous structural protein that makes up all connective tissue.
Collateral – A small side branch.
Communicated – The moving of an object from its proper place.
Computer tomography – An imaging technique that produces image slices through a patient which are then reconstructed by computer to form an overall picture.
Contraindicate – Act as an indication against the use of particular substances or treatment.
Coronary artery – The artery that supplies blood to the heart muscle.
Cranial – Pertaining to the superior end of the body.
Cut down procedure – Creation of a small incised puncture to enable entry via venous system.
Cysteine – A sulphur-containing amino acid.

D
Dentofacial surgery – Surgery involved with the teeth, alveolar, and configuration of the face.

Dermatitis – Inflammation of the skin.
Digital subtraction angliogram (DST) – A blood vessel imaging method in which the image is produced by subtracting background structures and enhancing the contrast of those areas of interest.
Distal – The farthest end from the reference point.
Dysphagia – Difficulty in swallowing.
Dyssynergia – Muscular incordination.

E

Embolism – The sudden blocking of an artery by an embolus that has been brought to its lodgement site by the blood system.
Embolus – A clot, usually part of a thrombus, that has migrated via the blood flow to lodge in a different region.
Endo – A prefix meaning situated within.
Endoscopy – A visual examination of internal structures using a specialized direct vision instrument.
Endothelial – Pertaining to the layer of cells that make up the inner lining of the heart cavities and blood vessels.
Esophageal – Pertaining to the passage connecting the throat to the stomach.

F

Face centre cubis – A crystal structure made up of cubic basic units. Each cube has an atom at each corner and an atom in the middle of each of its faces.
Femoral bone (femur) – The bone of the upper leg; thigh bone.
Fibroblasts – Immature fibre producing cell of connective tissue.
Fibrolytic therapy – A treatment that attempts to prevent fibrous tissue formation and thus reduce likelihood of embolization.
Flow dynamics – The assessment of turbulence created during fluid passage.
Frequency – Urination at short intervals without increase in daily volume of urinary output.

G

Gastroesophageal reflux – The return flow of the stomach contents into the esophagus.
Granulation tissue – New tissue formed due to damage of soft tissue.
Gross – Extensive.

H

Haemorrhage – The escape of blood from a ruptured blood vessel.
Hepatic – Pertaining to the liner.
Hexagonal close packed – A crystal structure in which the atoms form into a hexagonal base unit.
Histological examination – A study of the structure and reaction of tissues at microscopic level.

Hydrostatic pressure – The pressure exerted by a stationary fluid.
Hypercoagulability – Abnormally increased coagulability of the blood.
Hyperplasia – Abnormal increase in the volume of an organ or tissue due to the increase in the *number* of normal cells (production).
Hypertropy – Enlargement of an organ from an increase in the size of its cells.

I

Iliac artery (common) – The arteries to which the aorta splits up into so as to pass blood down into each leg.
Inferior vena cava – The vein that supplies the de-oxygenated blood from below the heart to the right atrium.
Infra – Beneath.
Infrainguinal – Below the groin.
Inguinal – Of the groin.
Intima – The innermost layer of a blood vessel.
Intimal – Pertaining to the innermost layer of a blood vessel.
Intraluminal – Within the lumen.
Intravascular – Within a vessel or vessels.
In vitro – Without use of a living body, i.e. within an artificial environment.
In vivo – With the use of a living body, i.e. within a living organism.
Isotropic – Affects the force of muscular contractions.

J

Jaundice – A morbid condition caused by an obstruction of the bile and marked by yellowness of the skin, constipation, lack of appetite and weakness.
Jugular – Pertaining to the neck.

K

Kyphtotic – Abnormal anterial – posterial curvature of the spine, also called hunchback.

L

Laminar flow – Non-turbulent movement of liquid nominally within a cylindrical vessel.
Laparoscopy – Examination of the abdomen and pelvic cavities.
Lateral – Pertaining to a side.
Lesion – Any discontinuity of tissue, either pathological or traumatic.
Ligation – The tying off of a blood vessel by any material.
Lingual – Pertaining to the tongue
Lumen – The channel within a vessel.

M

Magnetic resonance imaging – A non invasive imaging technique, utilizing the resonance of hydrogen atoms within structures.
Malignant – Tending to invade normal tissue and recur after removal (of tumour) – very virulent or infectious (disease).

Malunion – The faulty alignment of fractured bone.
Martensite – A metallic structural phase consisting of an irregular atomic pattern.
Metacarpus – The part of the hand between the wrist and the fingers.
Metatarsus – The part of the foot between ankle and the toes.
MGH – Massachusetts General Hospital, Boston.
Morbidity – The condition of being diseased.
Morphological – The science of the form and structure of organisms.
Mortality – The death rate.

N

Neo – New, recent.
Neoplastic – A new formation of tissue in some part of the body, i.e. a tumour.
Neurological – Concerned with the nervous system.
Neutron activation analysis – Examination technique using the activation of neutrons within the body to display specific areas of interest.
Nocturia – Excessive urination at night.
Non-union – When there is no alignment between bones.

O

Occlusion – The closing of a passage.
Oedema – Swelling.
Olecranon – The bony extension of the outside forearm bone (ulna) at the elbow.
Orthodontics – The branch of dentistry interested in the irregularities and misalignment of teeth.
Orthopaedics – The branch of medicine interested in preserving and restoring the musculoskeletal system.
Orthosis – An orthopaedic appliance or apparatus used to support, align, prevent or correct deformities or improve function or moveable parts of the body.
Osteotomy – The incision or cross-section cut of a bone.
Osterial tissue – The initial tissue which forms the basis of new bone.

P

Palliative – Serving to relieve (disease) superficially or temporarily or mitigate (pain).
Patella – The kneecap.
Patency – The condition of being wide open.
Percutaneous – Any act performed through the skin.
Perioperative – The time surrounding a surgical procedure.
Perorally – Through the mouth.
Phalanx – The throat.

Phlebothrombosis – The development of thrombosis without the formation of inflammation of the vessel wall.
Physiological – Pertaining to the treatment of a living organism.
Plastic deformation – A length or other physical change that remains after the applied forces have been released.
Platelet – A small disc-like structure formed in the blood.
Plexigas chamber – A special plastic tube used to simulate the outer vessel wall of an inferior vena cava.
Popliteal – The area around the knee.
Prophylactic – To attempt to prevent disease.
Prostate – Referring to the glands surrounding the neck of the bladder in male mammals.
Prosthesis – An artificial substitute for a damaged or missing part.
Proximal – The point on the subject area nearest the point of reference.
Pseudo – A prefix referring to something being false.
Pulmonary artery – The artery that takes venous blood from the heart to the lungs so that it can be re-oxygenated.
Pulmonary embolism – The blocking of the pulmonary artery by a blood clot.

Q
Quadriplegia – A condition in which the spinal cord is so damaged or impaired that all four limbs of the subject are in paralysis.

R
Radiological examination – An examination of the region of interest involving X-rays.
Rectal – Pertaining to the rectum.
Renal – The region around the kidneys.
Restenosis – Recurrent stenosis.
Rheumatoid arthritis – A disorder that causes painful inflammation of joints.
Roentgenographic – The taking of internal pictures using X-rays.
RSNA – Radiological Society of North America.

S
Saline solution – Salty solution in purified water that has very similar properties to normal blood.
Saphenous vein – A large vein originating in the foot and running up the inside of the leg to the thigh.
Sequela – The formation of a disease caused by some other event or condition.
Serum – The clear portion of the blood that does not contain fibrinogen or blood cells.
Sham operation – A fake operation in which the subject is treated in the same way as the usual operative procedure but no actual incision or implant placement occurs.

Sphincterotomy – Incision of a sphincter.
Stasis – The stoppage of fluid flow.
Stenosis – The narrowing of a lumen or opening.
Stent – A device that maintains a clear pathway by holding up the walls of the vessel.
Stoichiometric – Equal proportions.
Strain – The ratio of change in length (after an applied force) to the material's original length.
Stress – The ratio of applied force to cross-sectional area of subject material.
Strictures – An abnormal narrowing of a passage or duct.
Stroma – The basic framework of tissue on which functional tissue grows.
Subcubital artery – An artery just below the elbow.
Subcutaneous – Beneath the skin.
Superelasticity – An exceptionally strong ability to reform the material's original shape after a large force has been applied and released.
Surface arthroplasty – The placement of an artificial covering on the head of the femur to regain mobility within the joint.
Suture – A stitch or application of stitches or the material of a stitch used to close up an opening.
Symptomatic – To show symptoms of the subject complication.

T

Tetraplegia – A condition in which the spinal cord is so damaged or impaired that all four limbs of the subject are in paralysis, often referred to a quadriplegic patient who is ventilator-dependent.
Thrombosis – Formation of development of a thrombus.
Thrombus – A solid body attached at point of formation to the blood vessel and made up of various blood components.
Tracheobronchial – Pertaining to the air passages from throat to lungs.
Tracheomalasia – The softening of the air pathway.
Trans – Through.
Trauma – A wound or injury, normally considered to be from an outside source.
Triple strand twistflex – Three orthodontic wires twisted together to form a functional dental arch wire.
Tryptophan – A naturally occurring amino acid.

U

Ureteral – Pertaining to the tube which carries urine from the kidney to the bladder.
Urgency – Sudden compelling desire to urinate.

V

Vasodilator – Causing dilation of a blood vessel.

Vein – A vessel that carries used (de-oxygenated) blood back towards the heart.
Vena cava – Large vein supplying de-oxygenated blood from the body to the right side of the heart.
Venotomy – Incision of a vein.
Venous – Pertaining to a vein.
Vertebrae – A group of the separate segments (vertebra) of the spine.
Voiding – Act of casting out waste matter.

Index

Amino acids   12
Amplatz spider filter   40, 54
Anodic polarization technique   12
Antecubital vein   50
Aortic aneurysms   128
Arch wires   17, 18, 107, 134, 142
Arterial stents   1
Arteriovenous shunts   30
Arthrodesis   93
Arthroplastic cups   142
Arthroplasty   93
Austensite crystal structure   1, 2

Balloon expandable stents   71
Benign esophageal strictures   15
Benign prostatic hyperplasia   114, 117, 119
Biliary stenting   13, 111
Bilopancreatic diseases, management of   123
Biocompatability   5, 73
  of Ni–Ti alloy   5
Bird's nest filter (BNF)   40, 47, 52, 54, 56
Bladder catherization   119
Bracing technique   97

Cardiovascular endoprosthesis investigations   114
Carotid aneurysm   129
Caudal 'jump'   44, 49, 52, 53
Clots   27
  trapping   39
Clover leaf (mesh design)   34
Cock-up splint   101
Compressive staples   103

Contact dermatitis   11
Correcting spinal deformity   99
Corrosion behaviour of Ni–Ti alloy   5
Corrosion   11
Cranial aneurysm   128
Cranial clips   128
Cysteine   12

Deep venous thrombosis   50
Deformity of the spine   96
Dental arch wires   1
Dentofacial surgery   93
Distraction (Harrington) rod   97
Distraction rod technique   97
Dysphagia   13–15
Dyssynergia   119

Eccentric position   40
Eczema   11
Effect of stent geometry   78
Elastic flexibility   121
Elasticity   128
Emboli   27, 28, 34, 36, 39
Embolism pathway   27
Embolization   80
Embolus capture   39
Endoscopic sphincterotomy (EST)   123
Endothelialization   74, 76
Esophageal:
  lumen   13
  malignant strictures   13
  stenting   13, 113
External jugular   44

Fertility   119
Filter retrievability   62
Foreign bodies:
  retrieval of   124
  removal snares   134
Fractured tubular bones   99
Fractures   93

Gall bladder   123
Gap measurement   91
Gooseneck snare   125, 126, 128
Greenfield filter   54, 56
Gunther basket filter   40, 54
Gunther filter   54

Harrington rods   103
Histological investigation   100
Hypercoagulability   50

In situ saphenous bypass   84
In vitro corrosion tests   11
Inferior vena cava (IVC) filter   8, 34, 47, 54, 62, 63, 141
Instent   81
Internal jugular   42
Intraluminal disease   111
Intravascular endoprosthesis (IVEP)   69, 87
Intravascular stents   69
Intro-biliary tree stents   17
Introvascular grafts   132

Kimray–Greenfield (KG) filter   31–34, 36, 39, 62
Kyphtotic scoliosis   99

Laparoscopic operations   121
Laparoscopic spatula   122
Laparoscopic techniques   134
LGM filter   54
Ligation   30
Loop techniques   107
Lower jaw fractures   91
Luminal surface of Ni–Ti implant   73

Malignancy   50
Malocclusions   109

Martensite crystal structure   1, 3
Martensite transformation   2
Memokath   118, 119
Mesh designs   34
Migration   32, 47, 62, 118
Minimal access surgery (MAS)   121
Misplacement   62
Mitosis   11
Mobin–Uddin (MU) filter   29–31, 36, 39
Mock circulatory loop (MCL)   74

Neointimal thickness   76
Neuropathic bladder   119
Neutron activity analysis   7
Ni–Ti alloy filter   36
Ni–Ti catheter   86
Ni–Ti cups   93
Ni–Ti memokath endoprosthesis   118
Ni–Ti staples   93
Nitinol basket   123
Nitinol intratracheal stent   111

Oedema   30
Orthodontics   13
Orthopaedic bone implants   1
Orthopaedic clamps   103
Orthopaedic cups   103
Orthopaedic staples   103, 142
Osteosynthesis evaluation   91
Osteotomy   93, 94
Osterial tissue growth   94
Overlapping ring pattern mesh design   34

Percutaneous transluminal angioplasty   70, 87
Peroral cholangiopancreatoscopy (PCPS)   123
Phlebitis   30
Phlebothrombosis   29
Photoelastic test   91
Pitting   11
Prevention of pulmonary embolisms   141

Pulmonary emboli 47, 62
Pulmonary embolism 27–29, 40, 50, 62
Puncture site thrombosis 62

Quadriplegic patients 101

Rectal stenting 111
Removal of calculi 123
Resistance to fracture 121
Restenosis 73, 81, 87
Retrievable nitinol IVC filter 62
Retrieval of foreign bodies 124
Reverse knuckle bender splint 101
Root canal file 121

Scoliosis 96–98, 103
Self-actuating fasteners 24
Self-erectable space structures 24
Self-expandable stents 71
Self-expanding (superelastic) nitinol stents 13, 15
Sequela 63
Shape memory:
   alloy intravascular stent 142
   aortic prosthesis (SAP) 133
   effect 1
   stents 72
Simon nitinol filter (SNF) 39–41, 54, 56, 63, 142
SMA endoprostheses 87
SMA filters 62
SMA inferior vena cava filter 44
SN IVC filter 50
Spasticity 101
Spinal curvature 97
Spinal deformity, correcting 99
Spiral mesh design 34
Spring rate 19, 20, 111
Stenosis 78
Stent geometry, effect of 78

Stenting 13
Stored energy 19
'Strecker' stents 13, 15
Subacute thrombosis 80
Superelasticity 107
   of Ni–Ti shape memory alloy 13
Surface arthroplasty 93
   cups 103

Thermal airway stents 134
Thermal arch wire 111
Thermally activated devices 24
Thermally activated tracheobronchial stents 111
Thrombi, formation of 31
Thrombosis 27, 32, 130
Titanium Greenfield filter 54, 56
Titanium 12
Tracheal stents 1
Tracheobronchial stenting 111
Transition temperature range (TTR) 2
Transluminal angioplasty 70
Transstenotic pressure gradient 80
Tryptophan 12

Umbrella filter 30
Unwanted intravascular bodies, retrieval of 124
Ureteric stent 128
Urethral stenting 111, 114, 118
Urethral strictures 114

Vascular endoprosthesis 111
Vascular stents 70
Vena cava:
   filters 1, 27
   ligation 28
Vena tech filter 54, 56
Venotomy 33, 35
Venous pressure changes 39